Classic Tractor Collectors

RESTORING AND PRESERVING FARM POWER FROM THE PAST

By John Harvey

About ASAE — The Society for engineering in agricultural, food, and biological systems

ASAE is a technical and professional organization of members committed to improving agriculture through engineering. Many of our 10,000 members in the United States, Canada, and more than 100 other countries are engineering professionals actively involved in designing the farm equipment that continues to help the world's farmers feed the growing population. We're proud of the triumphs of the agriculture and equipment industry. ASAE is dedicated to preserving and record of this progress for others. This book joins many other popular ASAE titles in recording the exciting developments in agricultural equipment history.

Classic Tractor Collectors
Restoring and Preserving Farm Power from the Past

Book and Cover Design: Bill Thompson
Project Manager: Melissa Carpenter

Library of Congress Card Number (LCCN) 94-078878
International Standard Book Number (ISBN) 0-929355-56-3

Printed in the U.S.A.

Dedicated to my parents who studied, taught, and loved American history.

Introduction

Welcome to the fascinating world of tractor collecting. These unique machines and their owners are an integral part of our American agricultural heritage, which includes the coming of the gasoline-powered farm tractor during the 20th century. We all share in the legacy of the Tractor Age.

There are people living today who started farming with horses. And at one time early in this century, magnificent horses and mules helped produce food and fiber for America.

But as the population and demand grew, the tractor was necessary to fill the growing needs of a hungry world, and to free farmers of back-breaking labor.

Infinite in size and shape, tractors were designed with the best technology of the time. Some, like the horses before them, were driven hard and put away wet . . . abandoned when a a better model came along.

But there's a sentimental attachment to an old tractor. It's hard to explain, but it's real. So the dads and grandads who had used some of those machines—their work partners and field companions all those years—looked back and felt compelled to save those memories. By now the once proud machines were rust buckets wasting away in fencerows; a chain saw was the first restoration tool required.

The trip from fencerow to tractor show is a painstaking one, as you'll learn from several of the blood-sweat-and tears stories that follow. The devotion grows deeper, the pride stronger as the work progresses. To the people who restore them, the tractors become part of the family.

Call it a hobby. Call it a sport. Call it an obsession. Call it whatever you want, but a beautifully-restored classic tractor is a piece of history.

This, then, is American agriculture at its finest. These people and their machines tilled the soil and produced the food that fed our families.

Now, they are restoring and preserving a piece of our agricultural past. Tractor collectors, we salute you and thank you.

John Harvey

Acknowledgements

I've learned that a book is like an iceberg: the part you actually see represents only a small fraction of the whole thing. Many people were kind to me and helped with my overall effort in putting this book together. I could not have completed it without them and this is my opportunity to acknowledge them and say, "much obliged."

Special thanks go to Todd Stockwell, an agricultural historian, who is the caretaker of an ancient Hart-Parr (see page 6). Others involved in this project include: Jack Gilluly, Anaconda, Montana; Larry Jones, Smithsonian Institution, Washington, D.C.; Doug Strawser, Oregon, Illinois; The Farmers Museum, Inc., Cooperstown, New York; Floyd County Historical Society, Charles City, Iowa; Lake Farmpark, Kirtland, Ohio; National Museum of American History, Smithsonian Institution, Washington, D.C.

A significant number of the tractor photographs and their owners were taken for the Classic Farm Tractors Calendar, a project I launched in the late 1980s while a public affairs manager with DuPont.

My thanks to William F. (Bill) Kirk, who heads up DuPont Agricultural Products, for permission to use these beautiful illustrations.

The photographer who's taken the pictures for this calendar through the years is Ralph Sanders, a master at his trade. You'll get to see him in action beginning on page 111. Not only is Ralph an excellent photographer, he knows tractors.

A number of other talented photographers contributed to this book with their photographs of tractors and collectors. Thanks to: Kent Clawson, Eric Crossan, Duane Dailey, Jenny Harvey, John Harvey, Ed Lallo, Dick Lee, Richard Lighthill, Jim Ritchie, Richard Sanders, Dick Seim, *The Indianapolis News*, *The Sunnyside* (Washington) *News*, Tim White.

It goes without saying that this book could not have been published without the cooperation of the tractor collectors and their families, whose stories I've shared on the following pages. There is no question about it, tractor collectors are some of the nicest, neatest people anywhere. They are helpful, want to share, and are generous with their time. They believe in the work

ethic—and practice it. They are upstanding people with family values. They take great pride in their restored tractors, and rightfully so.

A great deal of gratitude goes to my daughter, Jenny, who took my scribbled notes, heavily-edited pages, and messy manuscripts and made them near-perfect on Word-Perfect. She does it all as my strong right arm. My wife, Carol, encouraged and cajoled me, whatever was needed to get a moody writer moving. She is a pretty fair country writer herself, and I trust her judgment. She sort of acts as my managing editor. (Even the family dogs, Tiffany and Dewey, were faithful office companions.)

My long-time friend and confidant, Dick Lee, was a cheerleader from the first time I mentioned my idea for this book. He assisted me with words of encouragement, story leads, and photographs. Dick, now retired, was the Agricultural Editor at the University of Missouri-Columbia, and was my advisor when I was a green student majoring in agricultural journalism at Mizzou. Through the years I conferred with Dr. Lee on every job change, every new career opportu-

nity. He was always there for me, just as he was for hundreds of other students. He still is.

Other agricultural communicators that I've know through the years were especially supportive. Dick Seim, one of the best in the business, was a *Farm Journal* colleague. Larry Harper, editor of *Missouri Ruralist*, has championed the cause of tractors and numerous other ag-related "objets d'art." Jim Ritchie and Tim White are two other experienced ag communicators who helped.

Special thanks to Barb and Dave Kalsem who so graciously gave of their time during their busiest week of the year—they directed the huge Waukee (Iowa) Swap Meet.

Finally, the people I've worked with at the American Society of Agricultural Engineers (ASAE), Donna Hull, Melissa Carpenter, and Bill Thompson, have been wonderful to work with. First-time book writers like myself need lots of guidance and reassurance. I thank them for their assistance and cooperation—and for publishing my thoughts about tractor collectors.

J.H.

Table of Contents

Most Famous Hart-Parr in the World

Two brilliant young men left a legacy that includes old No. 3

The deeper U.S. agricultural historians such as Todd Stockwell dig, the more they marvel at the accomplishments of two young mechanical engineering students at the University of Wisconsin-Madison, who collaborated in 1896 on an award-winning thesis expounding on the applications of the gasoline engine. The students, Charles Hart and Charles Parr, had mentioned just about every conceivable future use of gasoline-powered engines in their thesis except, curiously enough, agricultural field work.

Stockwell, an agricultural technology education and interpretation specialist at Lake Farmpark, a farm/museum not far from Cleveland, has done an exhaustive amount of research on Hart and Parr and the tractors they built. He feels he knows why the men chose to exclude agricultural traction engines from their honors thesis.

"Their innovative thoughts," Stockwell notes, "had already drawn the ire of their professors, several of whom were world-renowned steam plant engineers." Also, this was the turn of the century. "The role of the horse was firmly entrenched in American life and only a few editors were so bold as to say that the gasoline-

One of the most famous of all tractors, this 1903 Hart-Parr 17-30 was built in Charles City, Iowa. It featured a 2-cylinder horizontal, 4-cycle engine with a 9-inch bore and 13-inch stroke. No spark plugs were used. A make-and-break ignition, powered by eight dry cells, had points inside combustion chamber. The engine was oil-cooled and lubricated by sight feed, drip oiler. Old No. 3 was used by its first owner, George H. Mitchell, for 17 years. It is now displayed at Lake Farmpark, Kirtland, Ohio.

powered automobile might one day partially alleviate the burdens of draft animals. No one was speculating that the horse would ever be replaced in the nation's farm fields by mechanical power sources of any kind," says Stockwell.

It was against this background that Hart and Parr built and sold two gasoline traction engines between 1901 and 1903

to satisfied customers in the Charles City, Iowa, area, Hart's hometown.

The year 1903 can be characterized as the year of the gas engine. Stockwell points to three milestones that were to change America, and each involved the successful use of the gasoline engine: (1) the first successful sustained, controlled, heavier-than-air flight by the Wright

Brothers; (2) three successful automobile crossings of the continental United States; and (3) the successful commercial sale of 16 agricultural gasoline traction engines capable of performing both belt and drawbar work by the Hart-Parr Company.

It was a giant leap forward for agriculture. "The success of the Hart-Parr gasoline traction engines at both stationery belt and heavy drawbar work in the field (plowing) and on the road, offered a way to escape some of the technical difficulties encountered by steam engines on the western prairies, including lack of fuel, water, and the danger of fire. Likewise, the commercial success of this new independent manufacturer forced the established farm machinery firms to reevaluate the merits of the gasoline engine and ultimately led to the modern tractor," Stockwell emphasizes.

"Hart-Parr quickly became recognized as the founder of the gasoline tractor industry," he says, "and the year 1903 can be characterized as the year of the gasoline engine, a watershed of technological evolution that eventually resulted in profound social and economic changes

Todd Stockwell poses with Old No. 3 and a trio of young, interested tractor buffs. Stockwell is an agricultural historian and an authority on Charles W. Hart and Charles H. Parr and their early Hart-Parr tractors. Also pictured are Kevin Tompkins, Becky Sluss, and Lori Ward.

Old No. 3 gets lots of attention—as well a tender, loving care. Brand new, it sold for $1,580 on August 5, 1903. Here, Todd Stockwell of the Lake Farmpark staff, adds some oil under the careful scrutiny of Rachid Sigurdard and Colin Martin.

through the widespread adoption of the airplane, the automobile, and the tractor."

The basic 1903 design was used for tractors sold around the globe, until World War I caused the reorganization of the company and a subsequent shift to a lighter tractor. "Unlike previous gasoline traction engines, the Hart-Parrs were designed with a clutch and gear train capable of withstanding the strains of heavy drawbar work such as plowing," Stockwell notes.

International Harvester Company, with its vast resources, muscled past Hart-Parr in 1908 in total gasoline tractor production. This marked the beginning of the end of Hart-Parr's pioneering leadership in the field, but Hart-Parr had secured its place in agricultural history as founder of the tractor industry.

Of those grand old tractors built by Hart-Parr in 1903, only one remains, Old No. 3.

This priceless relic receives tender, loving care from Todd Stockwell at Lake Farmpark at Kirtland, Ohio, where it is on loan from the Smithsonian Institution in Washington, D.C. Stockwell has a world of facts and figures about Old No. 3, including a detailed chronology. A sampling follows.

August 5, 1903. Gasoline traction engine No. 1207 sold to George H. Mitchell, 1 mile south of Charles City, Iowa, on the Clarksville Road, for $1,580.

March 6, 1919. Testimonial letter from Mitchell published in Hart-Parr advertisement stating, "The tractor I bought of you in 1903 is still in good running condition after having been used each year since purchased."

July 8, 1924. Hart-Parr Company purchased No. 1207 back from Mitchell for $72.75 ($.50 per hundredweight).

January 18, 1932. Letter from Bert C. King, publicity manager for Oliver Farm Equipment Sale Company, Chicago, sent to J. W. Block, registrar of Museum of Science and Industry, Chicago, informing the museum that No. 3 and a new model

This drawing from an early advertisement shows the canopy; one is to be added to Old No. 3 to make it complete.

18-28 would be shipped from Charles City by railroad boxcar.

June 18, 1938. Letter from William S. Stinson, advertising manager of Oliver Farm Equipment Sales Company, Chicago, sent to Russell H. Anderson, Museum of Science and Industry, Chicago, requesting loan of No. 3 to the Omaha branch of Oliver for exhibit at the Iowa Centennial State Fair in Des Moines. Stinson suggested the museum ship No. 3 by truck to R. D. Merrill at the Des Moines Oliver branch.

June 30, 1948. Centennial year celebration held at Battle Creek, Michigan, by the Oliver Corporation featured No. 3 flanked by new Standard 77 and 88 Oliver tractors.

January 20, 1949. Internal memo to Janet R. MacFarlane, Curator, The Farmers'

Museum, Inc., Cooperstown, New York stating, " There is shipment to be made from Mr. Arthur Peterson of Buchen Company, 400 West Madison Street, Chicago, Illinois. Mr. Peterson called us to offer the first gasoline driven (or powered) tractor used in the United States. It is evidently owned by the Oliver Company."

April 30, 1949. Presentation talk by R. D. Merrill, Oliver Corporation, at Farmers' Museum, Cooperstown, New York, recounting Old No. 3's historical significance and presenting "this tractor to the Farmers' Museum of Cooperstown as a permanent exhibit."

August 1, 1955. No. 3 loaned by Farmers' Museum to Michigan State University, East Lansing, Michigan.

March 10, 1960. Letter from vice president of the Oliver Corporation to the Smithsonian Institution formally transfers title of No. 3 to the Smithsonian. The Oliver Corporation requested that the machine be returned to the company if it were ever to be disposed of.

April 25, 1988. Letter from John W. Mitchell, son of George H. Mitchell, sent to the Floyd County Historical Society (Iowa) about his memories of Charles City and No. 3.

July 24, 1991. Letter sent from Catherine Page, assistant registrar, National Museum of American History, Smithsonian Institution, honoring a request from Lake Farmpark, Kirtland, Ohio, for loan of No. 3.

July 30, 1991. No. 3 removed from exhibit in Washington and trucked to Lake Farmpark.

Their John Deere Tractors Are Calendar Quality

Bob, Ken, and Bill Waits are all John Deere enthusiasts. But the nub of the story is how they got involved in restorations

Collecting classic tractors is contagious, and the fever often spreads among family members —father to son, for example.

In the Rushville, Indiana, community, Robert (Bob) Waits has been an advocate of John Deere collecting since the early 1970s. Kenneth (Ken) Waits wasn't bitten by the bug until nearly 10 years later,

This father-son pair is a dynamic duo when it comes to restoring Deeres. Kenneth, the father (on left) and Robert Waits have had their tractors appear on national calendars. The R is Ken's; the BR is Bob's.

and then he, too, became addicted to antique tractors.

The two are father and son, and together they farm 1,300 acres of corn and soybeans. But Bob is the son and Ken is the dad; the son got the father started restoring tractors!

There's another link in this family chain: when Ken got interested, so did his brother, William (Bill) Waits. You

Bob's wife, Pam, is proud of her husband's handsome John Deere, too. Tractor collecting is fast becoming a hobby for the entire family to enjoy.

Rushville's Pioneer Engineers Club, one of the oldest and largest clubs of its type in the United States.

He's also convinced it was the 1949 John Deere Model A his grandfather purchased new—and one of the two-cylinder tractors Bob grew up with—that made his blood run John Deere green.

"It might be that the 'buddy seat' on dad's tractor had something to do with it, too," Bob says. "A buddy seat was twice as wide as a regular tractor seat. A dealer would put one on a demonstrator model and have a potential customer sit next to him during the demonstration. It helped make the sale. Dad's tractor had a buddy seat and I'd ride with him for

might say that Bob "infected" his father and his uncle, and the three are very happy the hobby spread within the family.

All three are quality-minded and do exacting work. Bob's mint 1939 John Deere Model BR was chosen for the DuPont Company's 1993 Classic Farm Tractors calendar while Ken's 1953 John Deere Model R was selected for the Deere & Company calendar in 1989. Bob and Ken still take a lot of ribbing from

the coffee shop crowd about being "calendar boys."

Bob's interest in vintage Deere tractors traces to 1973. At about that time he was dating and falling in love with his wife-to-be. He also fell head-over-heels for a 1938 John Deere Model L, Deere's tiniest tractor, and wound up buying it. He fixed it up and showed it locally.

Bob is sure that what whetted his appetite for old iron was the local antique tractor show, sponsored by the hours. After a while I'd get sleepy. It's a wonder I didn't nod off and fall off," Bob says with a smile.

Bob and Ken have about a dozen restored Deeres, which include an entire family of Model R's: an unstyled and styled AR, BR, and R diesel. They are all standard tread, somewhat rare to the Midwest. "The Model BR was designed mainly for open field work in the Plains states, the Dakotas, and the Canadian prairies," explains Bob.

He knows the history of his tractors forward and backward. His 1939 John Deere Model BR was built October 10 in Waterloo, Iowa, and was shipped to Yorkton, Saskatchewan. Bob shares these

This John Deere BR was built in the fall of 1939 and was shipped from Waterloo, Iowa, to Yorkton, Saskatchewan, Canada. Bob Waits did a magnificent job of restoration.

additional facts:

- The R designates it as a standard tread. The BR itself was never styled. There was never any sheet metal with screen wire around the radiator to make it look more streamlined.
- In June 1938, Deere revised the B motor and increased the bore and stroke to 4 ½ x 5 ½, giving it a total of 175 cubic inches.
- This particular tractor was capable of pulling two 14-inch plow bottoms in normal conditions. The new motor provided 18 ½ horsepower on the belt, 16 ¼ on the drawbar. The BRs came on either 11 x 28 rear rubber, or 24-inch rubber. They all came on 550- x 16-inch front.

That's the happy smile of a proud owner on the face of Bob Waits. Bob actually got interested in collecting and restoring John Deere tractors before his father.

- The carburetor is a natural drafting type. The cooling system uses thermal siphoning. With this system, there's no need for a water pump or a thermostat. It also has two gauges, one for water temperature and one for oil pressure.
- The BR has a PTO shaft: there's a lever on the transmission case to turn it on. To run a belt pulley, the operator would put the tractor in neutral and then engage the clutch lever.
- A grand total of 12,514 BRs, BOs, and BIs were made during a 13-year production run. (The R signified regular; BO meant orchard and BI industrial).
- The BRs were all-fuel tractors, so they have two fuel tanks. The smaller tank on the back held a gallon of gas used in starting. The larger tank held 12 gallons for either kerosene or petroleum distillate. The petroleum distillate was cheap; Deere advertised that using the distillate rather than gasoline would save enough money over the lifetime of the tractor to more than pay for it.

The entire family belongs to the Pioneer Engineers Club and the Two-Cylinder Club. They are regular participants at the John Deere Two-Cylinder Club Expos in Iowa, where only the very best restored Deeres are exhibited.

Bill Waits has restored nine John Deere tractors including the 1950 Model G he started farming with. It was Bill who found Ken's Model R diesel that appeared on the Deere & Company calendar.

Being crop farmers frees up these Hoosier State tractor buffs for restoration chores during the cold months. It took Bob most of two winters to bring his BR back to life. When he bought the tractor, it was in sad shape and wouldn't run.

"We basically tore it down to the ground, started rebuilding it and brought it back to what it is today," Bob notes.

Does his wife, Pam, mind that he spends so much time in the shop working on tractors?

"Heck no," Bob smiles. "She always knows where I am."

His Treasure Is Prairie Gold

Roger Mohr is a Minneapolis-Moline collector, museum curator, editor, and model builder.

Spend five minutes with Roger Mohr and you feel as though you've known him all your life. The Vail, Iowa, farmer is outgoing, loquacious, and naturally gregarious. He enjoys people and loves tractors. Especially Minneapolis- Moline, "Minnie-Mo." Just mention that name and his eyes twinkle like a kid's on Christmas morning. When Roger was six years old, Minneapolis-Moline introduced its distinctive Prairie Gold paint. He was smitten by that bright, beautiful color—and still is.

Iowan Roger Mohr is a walking computer with a marvelous memory bank on Minneapolis-Moline tractors, equipment.

To say simply that Roger Mohr is an expert on the Minneapolis-Moline Power Equipment Company of days gone by is a gross understatement. Mohr has a world-class collection of Minneapolis-Moline tractors; he opened a 24- x 40-foot museum in 1994 chock-full of such M-M memorabilia as literature, advertisements, promotional items of all kinds, dealer signs and such. He and family members also have a going, growing business for their scale model toy tractors. And Roger is publisher and editor of a quarterly newsletter, *The MM Corresponder.*

Roger's celebration of Minneapolis-Moline tractors began at an early age.

"My dad was a Minneapolis-Moline dealer in a little town with the name Ute, Iowa. I was seven when the Minneapolis-Moline 'Comfortractor' came to town. This was the classiest tractor of its time—perhaps of all time.

"They didn't truck that particular tractor, they drove it. They drove it all over the U.S., right to the dealer, who would give people rides. My mother tells me I was too sick to go to school that day, but I didn't miss a ride in the Comfortractor. That was in 1938, to be exact," Roger recalls.

As a child, Roger followed his father

With his wife, Marie, Roger Mohr makes a point about his Minneapolis-Moline tractors to a customer at the 1993 National Farm Toy Show in Dyersville, Iowa.

around like a puppy. "I'd curl up in those big, new tires and take my naps in the shop," Roger remembers. He also learned to be a mechanic, from watching and working with his father.

As a youngster, he would scrape, scrub, and then paint neighbors' tractors, earning $10 per tractor. "I'd spend a week or more on each tractor, so my

Like a king on his throne, this 1939 Minneapolis-Moline UTU surveys the countryside in western Iowa, with the Roger Mohr farm in sight. This handsome M-M is one of many Mohr has meticulously restored.

"They ain't all that pretty in the beginning." Eugene Mohr, one of Roger's sons, begins the restoration task by sanding a 1956 model in the Mohr shop.

wages were certainly reasonable," he chuckles.

When he married and started farming he bought, naturally, a Minneapolis-Moline tractor, a 1946 model Z.

In 1969 his hometown of Vail held its centennial celebration. Mohr remembers some of the local farmers talking during the festivities about the new tractors with cabs. He told them he remembered a tractor with a cab, the Comfortractor, when he was only a kid.

"Several of the fellows challenged my statement, so I was bound and determined to find one of the Comfortractors, officially known as a 1938 Minneapolis-

Moline UDLX, or U Deluxe. Only 150 of them were built, and they were way ahead of their time in concept and style," he says.

He searched for several years before stumbling onto one about 75 miles from his home. "It had been sitting abandoned in a shop for 11 years and was in absolutely terrible condition," Mohr relates.

With his sons Eugene, Martin, and Gaylen, Mohr spent a year and a half restoring the UDLX.

"People would come by while we were working on it, look at the rusted mess, and ask if I though it would ever run again. I'd tell them it was a U and that darn right it would run.

"Sure enough, when we finally finished it, she ran like a Swiss watch. Everything on the Minneapolis-Moline tractors was so big and heavy, nothing ever really wore out," says Roger.

That was the beginning.

One day he was looking at an old M-M sales brochure showing 20 different models. "We decided right then and there we should have one of each. Right now, we have around 70 tractors." Most of his tractors are picture perfect; in fact, they have been used to illustrate a book about Minneapolis-Moline models.

There's another dimension to Mohr's devotion to Minneapolis- Moline, his love for the miniature version. Several years ago he attended a toy show and noted, to his chagrin, that very few models of Minneapolis-Moline were available, though there was a plentiful supply of green and red toy tractors.

Roger decided to build toy M-Ms himself, one of every tractor the manufacturer ever made, just so he'd have his own personal collection.

Remembers Roger, "I parked a tractor in the garage and I lived with it for a year or more. I measured every single part of that tractor so I could reproduce it precisely at an exact $\frac{1}{16}$ scale. Every detail is there, with a total of some 70 individual parts."

At the first toy show, his model was an instant hit, and he returned home with 86 orders. Because of the intense interest, he and his three sons entered the toy model business, and the rest, as they say, is history. The mini-tractor parts are spincast in zinc. The individual parts are assembled, then carefully painted and detailed. So far, the Mohrs have produced 32 different Minneapolis-Moline models.

Their model business is doing extremely well: they can't keep up with the demand for their exacting miniatures. One of their big events each year is the National Farm Toy Show in Dyersville, Iowa, which attracts some 20,000 toy tractor buffs.

Roger edits and publishes *The MM*

Martin Mohr works on a 1949 2A, one of 32 different Minneapolis-Moline model tractors made by the Mohrs.

Corresponder, an informative newsletter "dedicated to the Great Minneapolis-Moline Power Implement Company," an informative newsletter from his farm near Vail, Iowa, for other "Minnie-Mo" enthusiasts.

Mohr publishes *The MM Corresponder*

with help from his wife, Marie, and son, Gaylen. Every article, advertisement, and illustration centers on M-M tractors, with news about upcoming conventions, a shoppers directory on parts, decals, paints and manuals, and letters from other collectors.

You'd better be up on tractor nomenclature if you shop the classified ads in Mohr's newsletter. Can you decipher the following?

FOR SALE: FE 339C + FE 340C (both pieces). Hot spot manifold set - $400. TY 170 spring - $5; Improved M5 starter - $170; KE 821C + KE 823 Hot spot manifold set — inquire (Contact for other needs).

The seller was from Pennsylvania, and just about any M-M hobbyist knows the advertiser was selling an exhaust manifold and exhaust damper and flange.

Like other vintage tractor restorers, Mohr is quick to share tricks of the trade with others involved in the sport. Freeing stuck engines is a common challenge, and Mohr discovered that olive oil works well, a trick he learned from a former Minneapolis-Moline mechanic. "You just pour the olive oil into the spark plug holes. It has tremendous penetrating power. Works better than kerosene," Mohr says.

One of his favorite tractors is the 1939 Minneapolis-Moline UTU, featured on the 1992 Classic Farm Tractors calendar.

"This was a universal type of a tractor—you could put a four-row cultivator on it; you could put a four-row planter on it; or you could use the tractor for plowing.

"They also made other versions of the tractor. The UTS had a standard front-end, more for wheat country. A UTI was an industrial-type tractor that manufacturing companies used around plants. And, of course, the tractor of all times was the Comfortractor, the UDLX. They're the same version, basically speaking—the same engine, same differential.

"The engine is a 283-cubic-inch, 4 ¼-inch bore, 5-inch stroke. It has a five-speed transmission, with Bendix brakes. A main feature at the time was the Visionline design. The rear end of the gas tank is very narrow, so visibility cultivating or planting corn was excellent," Mohr notes.

Optional features included lights, starter, and generator with dash panel. The belt pulley and power lift were also optional. The power lift was so designed that by depressing the pedal when the tractor was in motion, caused it to roll over and pick up the cultivator or planter. Mechanically, there were no hand lift levers.

"The grille is a red nose or red face. On the very first models the outside edges were very crude, there was no flat sheet metal. Little wires stuck out and it was very mean to put the grille in and out—it always scratched the hood," Mohr says.

The tractor came with a 5-speed transmission and a top speed of 25 miles per hour. "They had an exceptionally fast fifth gear in them," Mohr says. "The

Part of a sensational display of Minneapolis-Moline tractors and equipment in Roger Mohr's museum are the first and last M-M cab tractors—the famous UDLX and an A4T-1600 White Plainsman.

UDLX was advertised to have the capability of reaching 40 miles per hour. It wasn't quite that fast, but it was fast—especially for that time."

Mohr has come up with still another idea to keep the Minneapolis-Moline mystique alive. He's planning an annual open house for M-M collectors at his farm on Labor Day weekends. The three-day event will include a swap meet, and participants will be able to sell, camp, and tour his facility free of charge. He hosted a show for the 125th anniversary of Minneapolis-Moline on his Iowa farm in 1990. When Roger's first wife, Myrna, died suddenly in 1989, M-M collectors from throughout the country attended the funeral and showed their support for Roger and his family.

"Tractor collectors," says Roger, "are the salt of the earth . . . they are all sincere, honest people."

Call Him the Silver-King King

Living near Plymouth, Ohio, Paul Brecheisen has a passion for the locally built Silver King tractors

In the summer of 1991, Lake Farm-park, a 235-acre agricultural museum/park located 25 miles east of Cleveland, announced it was sponsoring an antique tractor restoration workshop, as part of its weekend-long show dedicated to antique tractors.

The workshop was a first-of-its-kind event, designed for novices as well as more experienced tractor enthusiasts. It covered the state of the tractor restoration hobby; provided tips on finding and

Paul Brecheisen played a major role when the Silver King Tractor Club was launched.

This 1936 Silver King R-66 was built in Plymouth, Ohio, by the Fate-Root-Heath Company, and was restored by Paul Brecheisen. The three-wheeler was capable of hitting 25 miles per hour or faster.

selecting a tractor for restoration; and told where to get parts and decals and how to obtain information through collector networks.

This workshop was a natural, since part of Lake Farmpark's mission is to collect farm machinery that represents important technological milestones and to carefully restore it to operational condition, so that it can be used in demonstrations at the park.

Brecheisen's R-66 has a 4-cylinder Hercules engine; later models had Continental engines. Low center of gravity made them excellent for mowing steep banks and Silver King was popular with highway departments.

One of the presenters at the restoration workshop was Paul Brecheisen, of Helena, Ohio, who trucked his 1936 Silver King to the museum. The situation was tailor-made for Brecheisen because he loves to share information about his Silver King—or any Silver King tractor for that matter. This guy knows Silver Kings like Arnold Palmer knows golf.

Living near Plymouth, Ohio, where the Silver Kings were built from the early 1930s to the mid-1950s, stirred his interest in this particular tractor manufacturer, its products, and its history.

"The Fate-Root-Heath Company built locomotives, but when the Depression hit, the firm added a tractor line to diversify. The first tractors built by Fate-Root-Heath in 1932 and 1933 were called Plymouths. That name didn't fit very well with the people building the Plymouth automobile, but since Fate-Root-Heath had been in business longer, they won the lawsuit. In the end, though, the name was changed to Silver King and the tractors were painted silver and red. The Plymouth tractors were silver and blue," Paul recalls.

Besides their distinctive silver markings, these tractors were known for their exceptional speed: some claim they could reach 35 to 40 miles per hour on the highway.

The tale is told of a man back in the early days who lost his driver's license because of being inebriated. Since he couldn't drive his car, he drove his Silver King to town. It was faster than his old car anyway.

Brecheisen has 15 different Silver King tractors, and one Plymouth, but his pride and joy is the 1936 Silver King Model R-66 (the number designates the inches between the rear tires). A tricycle model, it featured a single front wheel that ran between the corn rows and made it an ideal tractor for cultivating. This model became the standard for Silver King, and most came with rubber tires, still something of a novelty.

The tractor did not, however, have an electric starter.

Brecheisen smiles, "It has what we call an 'armstrong' starter," as he inserts the crank, gives it a quick twist, and hears the engine fire. His R-66 has a four-cylinder Hercules engine. Later models had Continental engines, as did a four-wheel unit Silver King made.

Paul tells about the time his father bought a tractor at a farm auction during World War II that came equipped with steel wheels. "We have a lot of rock around here and it's not easy riding. My father was at a tire store one day and someone had traded in their old tires for a new set. My father had to go before the rationing board to get a priority to get those used tires and replace the steel wheels, but that sure saved a bunch of backaches. I have steel wheels that will fit my R-66, but I'm not about to put them on," he laughs.

The first challenge in tractor collecting is to find the kind you want. Paul freely admits he had some friends who worked on the Ohio Turnpike that were like bird dogs. They spotted the Silver King in a

Though modern in many respects, you cranked to start. Lights and horn were available. With optional bull and pinion gear sets, tractors could travel up to 45 miles per hour.

barnyard and alerted Paul. "I went to look at it and asked the owner about its condition. He told me his son tried to pull it to break the motor loose, and something happened to the clutch. He was right. When we took the motor out, there was nothing left of the clutch at all. They had torn the facing off both sides of it," Paul notes.

Luckily, he was able to purchase a new clutch.

Networking with other Silver King buffs is one way to get parts, or to put buyers together with sellers. The Silver King Tractor Club that Paul helped organize has done all of the above and more. One of the first meetings the club held featured a grandson of one of the original company founders, who showed an old 16-mm movie film of a fleet of Silver King tractors maintaining highway rights-of-way.

Questions about paint colors are common from people ready to restore a Silver King, so the tractor club was quick to establish a paint color code.

The club is 500 strong and includes members in Australia, Scotland, and Wales. It has fueled new friendships and some late-night telephone calls across all parts of the United States and Canada. One such call came from a Canadian who had seen Paul's Silver King on the Classic Farm Tractors calendar. He was fascinated—he didn't know there was such a tractor and wanted to hear about it from the owner.

Their conversation continued, and the man from Saskatchewan disclosed the fact he owned several Gibson tractors and might want to sell one. That really got Paul's attention. Paul's first love after the Silver King is the Gibson; they were marketed for only a few years in the early 1950s. Though Paul has one Gibson, he listened intently and gathered the details since the Canadian had a different model.

Later, Paul mentioned to his wife that he was pondering the idea of driving to Canada to inspect a tractor. She liked the idea; it sounded like a nice vacation. "Then I had to confess to her that I had looked at the map and it would be a long drive—maybe three days out and three days back . . . in a pickup . . . pulling a trailer."

Paul's Canadian trip is still on hold.

This Silver King was ideal for cultivating corn or other row crops as its single front wheel ran between the rows. Available on steel or rubber, most farmers wanted tires and Silver King is credited as one of the firms promoting tires—Allis-Chalmers was the leader.

A Tractor By Any Other Name Isn't Quite the Same

Marty Huber had no farm background, but now she's joined the hunt for a Huber

People who collect and restore old farm tractors are the nicest people on earth. If that sounds presumptuous, just remember what Baseball Hall of Fame pitcher Dizzy Dean once said: "If you can do it, it ain't braggin.'"

Martha (Marty) Huber discovered the preamble about tractor people was a truism when she first began accompa-

The names and emblems on their t-shirts designate their favorite tractor—the Huber, which was once built in Marion, Ohio. Martha (Marty) Huber and her husband Don aren't related to the man who founded the company, but they do enjoy finding a tractor with their name on it!

nying her husband to events centered on antique John Deere tractors. Marty came in contact with the wives and women at these gatherings.

"The ladies are so genuine, so friendly and easy-going. While the guys are talking and retalking tractors, we gals check the craft stalls, sit around in lawn

Marty Huber's Huber is a cardinal red 1943 Model HK, one of the last the company made. Her husband gave it to her on their 25th wedding anniversary.

This 1928 Huber 25-50 is owned by Don and Marty Huber, of Moline, Illinois, but the tractor was photographed on the family farm near Oxford, Wisconsin, where Don's mother and two brothers live. A big machine, it stands 5 ½ feet tall at the hood. Only 283 of these tractors were built. The 25-50 weighed more than 4 ½ tons.

Ornate trim and a handsome logo decorate the wide fender.

chairs, cross-stitch, and have great conversations. We talk about anything, a lot more than just tractors . . . these are very special people to me," Marty comments.

Marty's interest in tractor restoration and its people cranked up in the late 1980s when her husband, Don, was gathering information about the early John Deere two-cylinder models built from 1918 to 1960.

Don Huber, an advertising executive at Deere, is coauthor of the coffee-table book, *How Johnny Popper Replaced the Horse*, a well-written, beautifully-illustrated history of the two-cylinder tractors built by Deere & Company.

The more he learned, the more she learned. "My job at this point was to feed the author," Marty recalls, "because he wrote mostly at night and on weekends." The book was completed and published in 1988, and by then both Hubers knew that tractor collectors were going to become an important part of their lives.

Don's new friends kept berating him because he didn't have an antique John Deere tractor and also because he knew so many avid collectors. "Well, one day, a Plains City, Ohio, friend changed his approach," Don remembers. "All right," the friend said, "if you're not going to get a John Deere, then you need a Huber. And I just happened to see one this afternoon . . .."

"Don liked the idea of a Huber owning a Huber tractor," Marty explains, even though Don's genealogy does not trace to the Huber company founder. Edward Huber was a mechanical genius who designed and built farm equipment, steam engines, and tractors, and held more than 100 patents—though he had only 13 months of formal education.

"Although Don grew up on a Wisconsin farm, he wasn't acquainted with Hubers," Marty notes, "and since I was a city girl I had no knowledge of old tractors, period. He became fascinated by those half-century-old, and older, tractors that bore his name. No, our name—after all, it is my name, too. Huber tractors are not nearly so numerous as, for example, John Deere because only a few hundred of each model were made."

It wasn't long before the Hubers, of Moline, Illinois, owned their first Huber. Several more purchases followed in rapid succession.

Dick Bockwoldt, of Dixon, Iowa, helped them locate several. Bockwoldt, a master at tractor restoration, restored Don's 1928 Huber 25-50, the tractor that made the 1994 Classic Farm Tractors calendar. (For more on Dick and his passion for antique tractors, see the chapter entitled Dick Bockwoldt: Tractor Restorer Par Excellence.)

Solid, heavy Huber was built to last and last.

In all, Don and Marty have nearly 20 Hubers. Two are freshly restored and another was repainted a few years ago by the previous owner, and several will eventually be renovated. Others will be cannibalized for parts.

For their 24th wedding anniversary, Don gave Marty a radiator top from a Huber. "I'd told Don that when our children were out of the nest I wanted to get a red convertible. For our 25th wedding anniversary, he followed through with a red tractor, a 1943 Huber HK, one of the last the Marion, Ohio, company built. My tractor is older than me, but it's much more modern than Don's, because mine starts on a battery, has lights, and rubber tires," Marty smiles.

Most would agree that her brilliant red Huber outshines Don's tractor, painted a World War I army-olive color. The paint was left over from the war effort, so the Huber people painted tractors with it until it was used up, sometime in 1928—shortly after Don's tractor was built. (However, the bright orange wheels and hunter green trim do brighten up the drab color.)

Don's 25-50 is unique in other ways. Though called a 25-50, it was rated officially at the Nebraska test in 1929 as having 50 horsepower on the drawbar, and 69 ¾ horsepower on the belt. "They re-rated it, conservatively, and called it a 40-62. So, most people that are familiar with this size tractor would probably know it as a 40-62 that was painted blue-green with red wheels," he explains.

A four-cylinder, 618-cubic-inch Stearns engine powered the 9,000-pound tractor that stands 5 ½ feet tall at the hood. Just 283 were built. The Huber had two forward speeds; wide open it would do a whopping 3 ½ miles per hour.

Don and Marty plan to be a part of the 1995 celebration in Marion, Ohio, when a museum is opened to honor Edward Huber and the company he founded. Besides establishing himself and his products in agriculture, Huber financed the Marion Steam Shovel Company, which manufactured machines that helped carve out the Panama Canal.

They will also pencil in several major tractor shows on their calendars, such as the Midwest Threshers Reunion in Mt. Pleasant, Iowa, the granddaddy of tractor shows in those parts. Marty's tractor, also restored by Bockwoldt, won Best of Show honors in one outing last summer. She and Don both wear shirts that carry the Huber symbol and name, and a friend of Marty's painted each of their restored Hubers, in a farm setting, on a dress Marty wears to tractor shows and gatherings of collectors.

"When we have friends over to our house for a social event or cookout, we often park one of our Huber tractors on the driveway, so when our son got married, he was scared to death we'd drive a tractor to the church. We didn't, but we did think about it," Marty muses.

A four-cylinder, 618-cubic inch Stearns engine powered it.

Meet "Mr. Massey"

The past lives in the gleaming machines Keith Oltrogge owns and writes about

When dad bought a new tractor back in the 1940s and 1950s, a young son often got a new tractor, too. A toy tractor—same model as dad's—a gift from the dealer.

Imagine the shine in that kid's eyes. And did that imprint a particular make in the junior farmer's mind? With an investment of a buck or two, that nice gesture by the dealer earned him a lifelong friend. Keith Oltrogge, Denver, Iowa, accountant, farmer, and editor of *Wild Harvest*, a magazine for Massey collectors and friends, started his toy collection with dealer and family gifts. A Massey-Harris 44 pulled a Clipper combine in the make-believe fields of his childhood. He remembers his dad coming in dust-caked from the field off those same real units.

He's an accountant, farmer and an editor. First and foremost, Keith Oltrogge is a "Massey man."

Yes, he still owns those toys, two of 600 tractors and implements that line floor-to-ceiling shelves on the four walls of his "toy room." Other models followed the 44 as his father, Orville, traded or added equipment. Christmas or birthdays brought more.

As Keith grew older, he added to his collection. But after he received his accounting degree in 1973, he expanded his purchases to "big boy's toys." He bought a Massey-Harris 30 in 1975; then acquired a 1936 Challenger, Massey's first row-crop tractor.

Today his Massey passion has packed barns and sheds on the family farm. Keith owns close to 25 tractors, all but four or five of which run, plus a self-propelled corn picker and other implements. He uses them for farm or field chores from time to time.

"It's a lot of fun to work the antiques," he says. "It gives you a sense of your father's or grandfather's experiences in the years they used the equipment."

Keith could hardly avoid becoming a "Massey man." His grandfather, Herman, bought a Wallis (predecessor of Massey-Harris) Model OK back in the 1920s.; he added a Model 25 in 1930 for plowing and belt work.

In 1938, Herman brought home a 101 Jr.—his first Massey row-crop tractor. The workhorse for the 1940s and 1950s, a Massey-Harris 44 (which had 45 horse-

Wild Harvest is the magazine that functions as a nationwide exchange for Massey-Harris collectors. Keith has to hustle to hit six publishing deadlines a year.

power, pulled three 16s, cultivated four rows) came along next, soon joined by a Massey-Harris 30 (30 horsepower).

Keith still farms with Massey tractors. A 1980 Massey-Ferguson 2745 supplies

Oltrogge has a huge collection of Massey-Harris models, but that's just the tip of the iceberg. He has signs of all kinds, literature galore, even a Massey-Harris cream separator. He's actually a Massey archivist, as the company refers questions to him.

the main muscle; a 1975 Model 1105 provides backup. He pulls the corn planter with a 1975 Model 265 and hangs a loader on a 1978 Model 255.

The "one-sixty" farm of his early years has grown to 350 acres. Keith, his wife Susan, and children Christopher, Scott, and Kate moved back to the farm five years ago. His parents, Orville and Hazel, retired to town.

Keith grew up on land his great-great-grandfather William, Sr., settled in the 1840s, in a house William, Sr., built in 1871. It stands square and solid today; the original brick now coated with pebble-dash stucco. (Great-grandfather William, Jr., succeeded his father on the farm.)

That family continuity, and, Keith says, "a custom of never throwing anything away," nurtured a collector's instinct.

As youngsters, Keith and sister made a small shed their treasure house. They filled it with crockery, old tools, scythes, weather vanes, license plates, and countless small things forebears had saved "just in case."

Awakens memories, doesn't it? Keith, however, never stopped collecting: he has acquired tractors, gas engines, and yes—Massey-Harris cream separators! He's sought out catalogs, manuals, promotional and sales movies and literature—in short, anything and everything pertaining to Massey tractors and equipment.

A Massey dealer closing out? Keith scouts office areas, gleans pertinent papers from files, display racks, parts counters. He drifts through flea markets, swap meets, and antique shows, eyes alert for Massey materials.

At home and in his office basement, he has organized the materials and filled steel files. He has become not just a collector, but unofficial archivist of the Massey-Harris, Massey-Ferguson lines. The company often refers questions to him.

Keith's own interest and dedication to this subject made him the natural choice when the Massey collectors' magazine needed a new editor in 1986. He helped found the publication in 1984.

Circulation has grown from 250 to 1,000 under Keith's editorship. He'd like to push that to 1,500 or 2,000. "That would let me consider things like color on the cover," he says.

"We try to get them out on time," Keith says with a grin, "but it gets pretty hectic around here." His wife Susan, who works in his office, helps hustle *Wild Harvest* to press. Subscribers and collectors contribute articles, and Keith often draws on his voluminous library to background features he writes himself.

A typical issue numbers 24 pages. It includes articles about various Massey models, black and white photos, pages of classified ads, and a few display ads. Most ads deal with tractors or parts. The magazine thus functions as a nationwide exchange for collectors and also serves as the focus for Massey collectors, an informal organization without officers.

Many hobbyists know Keith as the single best authority on their "addiction." They go to the phone. Lynn Goos, Silver City, Iowa, farmer and collector, says, "He's really been great. When I call him for information, he'll talk at length. And he'll return calls.

"He helped me locate a specific model, provides reprints of Massey materials at

An example of the memorabilia from Keith's collection.

reasonable cost, makes an effort to stop by if he's in the area. He does an excellent job. It's nice to have someone to go to when you need help in this area."

McComas Albaugh, Union Bridge, Maryland, retired farmer and postman, echoes that praise. He also recalls a time when he and a friend found a way to return a favor. "We attended an auction at a dealership up in Pennsylvania—bought up all the literature. We shipped it to Keith. I think it came to 260 pounds!"

Albaugh, another committed Massey man, owns 70 to 75 tractors. "Don't remember exactly, I'd have to count," he says with a chuckle. He also hoards literature. "From the time I was a boy, I'd gather up leaflets at all the machinery displays at fairs. Still have 90 percent of it. It's looking like a good investment."

Frequently, devotees lack the "paper" for a treasured piece of machinery. Many turn to Keith. He usually can find it in his files and will supply an extra original or a copy. For example, a man called who needed a manual so that he could operate a recently acquired Massey-Harris No. 26 grain drill. Keith found the manual, copied it, and put the man in the field.

Keith harvests a little payback from his hobby in this and other ways: He records Massey field days and events with a video camera and sells copies. (He organized the annual "Massey Day" and other affairs with the help of local groups.)

He also employs a professional studio to transfer old dealer training and promotional films to video tape. "A master tape costs me $400 to $500. So at $20, I need to sell at least 20 copies to break even," he says.

In addition, he reproduces some printed materials and decorative decals for Massey tractors and implements to sell.

Keith's biggest rewards, however, come from working with the machines and memorabilia of a passing era—and fellow collectors who share that interest. For Keith, the issue goes deeper than a love for red and yellow paint and the smell of hot metal and oil. "These tractors represent the living history of your own farm, your own heritage. In that sense, they carry family values of generations past into the present.

"When I crank up an antique tractor, I bring to life memories of the times of my father, grandfather, and great-grandfathers; what they believed and what they stood for."

Always on the hunt for Massey-Harris material of any kind, Oltrogge frequents tractor shows and swap meets. Note his wristwatch—Massey-Harris, of course.

If Keith's life reflects those values, they included rock-solid integrity, hard work, devotion to family, and concern for others. That willingness to invest himself in others' interest profits his fellow collectors. Few would argue if we acclaim him "Mr. Massey."

A Long-Lasting Love Affair with Case

Herbert Wessel has a barn full of beauties and a heart full of love for Case tractors

Any evening the phone rings at Herbert and Mary Wessel's home near Hampstead, Maryland, there's apt to be a Case tractor enthusiast on the line calling from some far off corner of the country requesting information or wanting to share some news. Antique tractor collectors have their own hotlines and networks: they'd already built an information superhighway before it became popular. Herb Wessel is part and parcel of all this because of his knowledge and intense interest in Case tractors and equipment.

His Case collection includes more than 30 tractors, an early Case steam engine, and a sporty Case automobile. His sleek-looking 1918 Case 9-18 was the cover tractor for the 1994 Classic Farm Tractors calendar.

He's a Case collector—of anything and everything that bears the Case name or the Case eagle. He's Herb Wessel, of Hampstead, Maryland.

The first Case he saved and restored was his father's, a 1941 Case DC. However, it's just one of many showroom ready models parked in Herb's immaculate stable that elicit oohs and aahs from visitors. His collection is a representative sample of every tractor Case made from the beginning through the 1960s.

Memorabilia are everywhere. There are signs, banners, advertisements, literature, and specialty items promoting Case tractors and equipment. Saving equipment and machinery, incidentally, is one of his pet projects, and Wessel makes a convincing argument for preserving these pieces—now.

He's certainly doing his part. "I have several barns in which I store various kinds of equipment, and I'm constantly looking for rare items, especially if they are partly wood. If not protected, the wood rots and we've lost a part of our farm history," he says. "A binder doesn't last long in a hedgerow."

An example of a recent acquisition is a Galloway manure spreader built in the early part of this century. Its wheels are wooden. (He also possesses a rare Galloway tractor, made in Waterloo, Iowa, that's waiting patiently to be restored).

One of his favorite shopping grounds is a huge auction in Lancaster County, Pennsylvania, an annual February gathering, where there's a plethora of antique farm equipment, much of it horse-drawn and of particular interest to the community's Amish farmers.

"Here's where I found a Bennett two-row corn cutter. The machine cuts the corn and lays it on a platform to be picked up by hand and set on a shock. It was pulled with one horse and was made in Ohio in the early 1900s. It's a rare piece, the kind I'm always looking for and hope to buy—providing the cost isn't exorbitant," he says with a wink.

Herb encourages tractor collectors to exhibit a piece of machinery with their tractors whenever possible. "I feel that Case collectors are probably doing more than others in this area. If we all exhibit a tractor that has a plow or some piece of equipment, it will help educate those who are totally unfamiliar with farming, and there are more of those people every year," he points out. "Let's begin right now to preserve and exhibit farm equipment, along with our restored tractors," Herb says emphatically.

From the first, Wessel was interested in anything mechanical. He has a deep appreciation for antiques and American

This 1918 Case 9-18 came equipped with a 4-cylinder cross-mounted engine, and two forward speeds of 2 ¼ and 3 ½ mph. Owned by avid Case collector Herbert Wessel, it is one of an entire fleet of tractors he has, a representative sample of every model Case built from the beginning through the 1960s. He also has a rare Case automobile, and numerous pieces of farm equipment.

history, and participates annually in the Maryland Steam Historical Society Show as well as the Mason-Dixon Historical Society Show.

Herb knows his Case cars, too. "Case got into the car business when it acquired the Pierce Racine Car Company—not related to Pierce Arrow—in 1910. Case built cars until 1927, though production was somewhat limited; fewer than 30,000 in total. In 1925, production was about 1,000 cars, of which 139 were the Model X like the one I have. Fewer than a hundred Case cars of all models and years are known to survive today.

"Case thought every farmer would buy a Case automobile because of the excellent reputation it had established with its steam engines and tractors. However, a Case car sold for more than $2,000, while a Model T Ford cost only about $400. Case autos were hand-built, not built on an assembly line like a Ford," he points out.

His sleek 1925 Case Model X Suburban Coupe is so good-looking it makes your mouth water. The only one known to exist today, Wessel bought the one-of-a-kind car at an estate sale in 1988. He completely restored it.

Says Herb, "I traced the car back to its original owner, a Mr. G. B. Gunlogson, of North Dakota. I am the fourth owner. The car was in reasonably good condition. However, its 55-horsepower, six-cylinder Continental engine needed work, and the interior of the car was in bad shape."

This is a classy automobile. In the

1920s, it competed with Cadillacs, Packards, and Pierce Arrows, all of which were known as expensive, large, top-of-the-line cars.

Sneak a peek at Herb's Model X and you'll note 30-inch doors for easy access from either side; front wheel spindles that turned on ball bearings for easy steering; heavily upholstered seats for comfortable riding. The driver's seat was permanent, but the passenger seat tilted forward, making the rear of the car easy to enter. At friction points, antisqueak materials were used. The car had a rear-mounted gas tank with a vacuum fuel system. Dried ash lumber was used in the framework, reinforced with malleable iron braces. Its Goodyear Cord tires were 32 x 4 ½-inch size—white sidewalls, of course. Adding further smartness to this sporty model were polished aluminum luggage guards and aluminum scuff plates.

When Wessel was asked to select his favorite Case tractor for the Classic Farm Tractors calendar, it was almost a mission

impossible, but he finally settled on the 1918 Case 9-18.

"This was the first small tractor Case built. Up until that time Case and the other companies built the huge models that broke the prairies and powered the threshing machines and saw mills. But [around 1915], Case thought they should come out with a tractor to try to replace the horses. This was the Case Company's answer.

"It was a pretty nifty tractor in its day, with a lot of sheet metal, fancy paint, and striping, which I think they probably thought the farmers were going to go for and would help sell the tractor. They built about 5,500 of these tractors in the three years that they were out. This

It looks good either way you view it. This model first appeared in 1916 and caused quite a ruckus with its bright green body, red wheels, and the fancy striping and lettering.

particular one was built in 1918. Later they changed the design to a 9-18 B, which was a little bit more practical without all the sheet metal. "Being the first small tractor that the small farmer could use was important. And, even though it had some design problems, it was probably necessary to come out with something like this to gradually evolve into what was to come later in small tractor lines," Herb says.

The first version of the 9-18 was strictly for gasoline; later, they made a kerosene model. The tractor has a four-cylinder engine mounted cross-ways, with a 3⅞-inch bore and 5-inch stroke. It runs at 900 revolutions per minute and has 9horsepower on the drawbar and 18 on the belt

pulley.

Even though this was a small tractor, Case claimed it would pull a two-bottom 14-inch plow, or a 20- x 28-inch threshing machine. This tractor has a transmission with two speeds forward, one reverse. Another interesting feature about this tractor is the round, spoked wheels, apparently a carryover from the steam engines. Most of the other Case tractors had a flat, spoked wheel, and only the 9-18 and the 10-20, both of which were early Case tractors, had the round spoke.

"Another unusual thing is the differential lock. If you get in a tight place where one wheel would be doing all the spinning, you could actually lock the differential together on these wheels to pull you out."

The cross-motor design is worth noting, too. Case used it as a sales

feature. They claimed that because the engine was set crossways, all of the shafts would turn in the same direction all the way back to the wheels. Therefore a lot of parts, namely, beveled gears to change the direction of the wheels were eliminated. "Of course, in 1928 when they changed to regular engine mounting, they had to eat their words," Herb grins.

Herb and his wife, Mary, have two sons, a daughter, and six grandchildren. One son farms with Herb, the other restores tractors with him." Two of the grandchildren are crazy about the tractors and insist on sitting on every tractor seat," Mrs. Wessel says. His lineup of Case tractors underscores the fact that in choosing its colors for its tractors, Case was all over the rainbow. "There was green, dark gray, light-gray, Flambeau red, desert sunset, red, and black," notes Herb.

Herb usually restores a tractor a year. He has nearly completed his set of Model R Case tractors, which include a 1935 RC, 1937 RC, 1938 unstyled R, and 1939 RC styled. These models have a horizontal, distinctive "sunburst" grill, an avant-garde design.

A stickler on safety, he encourages all antique tractor owners to carry a fire extinguisher on their equipment, "just in case there should happen to be a fire."

Herb also recommends wheel chocks to place under the wheels when tractors are parked, especially if the terrain is a little hilly. "An accident would take all the fun out of a show, and we don't want that."

He Revived a Sleeping Giant

With dedication and perseverance, Irvin King rescued and restored a 1910 Pioneer 30-60

South Dakota's searing summer sun and its numbing winter cold had surrounded the old tractor's carcass for more than half a century, 55 years to be precise, when Irvin King found it hidden behind a cattle wind break at the edge of a tree belt near Ree Heights.

Its massive rear steel wheels had been stripped years earlier and converted into stock watering tanks so strong not even the meanest bull could cause them damage. The cab had rotted off, and steering parts, the radiator, the drawbar, and front wheels were missing, salvaged for strap iron.

That's how the 1910 Pioneer 30-60 looked that day in 1980 when King discovered it. But you should see it today!

This behemoth of the prairies, built when this century was only 10 years young and William Howard Taft was president of the United States, has been completely restored by King and has appeared on the Classic Farm Tractors calendar, been featured in magazines, and appears on tractor trading cards.

Incredibly, the old tractor's innards were in good shape, all things consid-

It was a herculean undertaking on a gigantic tractor, a massive machine that ruled the prairie early in this century. Irv King to the rescue.

ered, mostly because of the heavy tinwork that protected the motor. "The engine was only partly stuck, freed up easily, and I didn't even have to clean the carburetor to get the engine started," Irvin recalls. "It was truly amazing."

Irvin knew one of the original owners, Edgar Knox, and consulted with him regarding the tractor's correct colors and

numerous other details. The elderly Mr. Knox still had the original owner and parts manuals, correspondence with the factory in Minnesota where the Pioneer was built, even some pictures taken of the tractor in 1913 on the Knox Brothers Ranch of Edgar and Harry Knox. This proved to be a bonanza because there was precious little information anywhere about the company or its tractors. Irvin met with Mr. Knox several times before the latter died at the age of 93 in 1982.

This combination of tenacity, dogged detective work, and good luck led to the preservation and restoration of this Pioneer tractor. Its KW magneto was found at the farm where the Knox brothers had stored it when the tractor was parked and abandoned. Irvin located rear rims and front wheels in Montana. Some of the original rear spokes were still with the tractor, providing a pattern for new ones. Irvin repaired the water pump and friction drive and replaced the cab and missing pieces of the hood. He did it all in his farm shop near Artesian, South Dakota.

"This tractor was the seventh one built by Pioneer in its Tractor Works in Winona, Minnesota. Since the company

This is an awesome tractor, folks, and it was way ahead of its time. The 1910 Pioneer 30 was faster than other tractors—it would go 6 mph in high gear. It boosted a three-speed transmission, a totally enclosed cab with an upholstered seat, and a drilled crank-shaft with forced-fed lubrication—all firsts on the market at that time. It weighed 12 tons and cost $3,000. Its four-cylinder horizontally opposed engine had a 7-inch bore and 8-inch stroke. Quite a tractor.

Its 1,232-cubic-inch, horizontally opposed four-cylinder engine, with a 7- x 6-inch stroke and bore, ran as smoothly as a Singer sewing machine, with little vibration. In a period when giant, gasoline-powered tractors battled steamers for bragging rights in the countryside, this tractor was far ahead of its time.

"It was the first tractor on the market to sport a three-speed transmission and was faster than any other tractor at 6 miles per hour in high gear. This tractor had a totally enclosed cab with an upholstered seat, glass windows, and a back curtain. A drilled crankshaft featuring force-fed lubrication was another first-on-the-market feature," Irvin notes. (It did, however, have a chain steering operation, like a steam engine).

The engine had an expanding shoe clutch in the flywheel. The carburetor was positioned at the bottom of the engine so the kerosene was sucked up through two cast iron tubes built inside the engine crankcase to preheat the kerosene for better ignition, since the hot oil was always on top of the cast-iron tube keeping it warm.

The manufacturer, Pioneer Tractor Company, offered a guarantee on its tractor, another first in the fledgling tractor business.

This is a big tractor. The rear wheels are 8 feet tall, 3 feet across the face (with 12-inch extensions). The rear of the tractor is 12 feet, 2 inches wide. From

was primarily building construction equipment, they specialized in large tractors. The Pioneer 30-60 was a diversified tractor in that it could be used for road construction or for farming. It developed 30 horsepower on the drawbar and 60 on the belt—this at 600 revolutions per minute," Irvin says.

The engine used 50 to 60 gallons of kerosene and a like amount of water in a day. The fuel tank held 100 gallons.

Irv King stands behind a front wheel that's 5-feet, 5-inches high. THe rear wheels are even larger, measuring 8 feet from the ground to the top. King, of South Dakota, located the rear rims, front spindles and front wheels he needed for the restoration in Montana. Irv spoke to the original owner, Mr. Edgar Knox, several times before his death in 1982 at age 93.

ground level to the top of the cab it's 11 feet, 2 inches; the tractor is 19 feet long. Front wheels are 5 feet in diameter, with a 12-inch facing. It tips the scales at 26,000 pounds.

It was this cumbersome size that eventually doomed it and other large gas-powered tractors in those early days.

Irvin learned that the Knox brothers ordered the tractor at the 1910 South Dakota State Fair, and it was delivered by rail from the Minnesota factory to Huron, South Dakota, the rail siding closest to their farm, some 24 miles south. It was priced at $100 per drawbar horsepower, so this tractor listed at $3,000. Edgar and Harry Knox

paid $2,850, plus $75 for shipment. Later they bought extension rims for better traction.

Irvin started playing and collecting old gas engines at around eight years of age. He always wanted an old tractor and in 1968 found his first tractor, a 1919 Pioneer Special, which is more rare than the 30-60. The tractor bug hit hard, and Irvin and his son, Mark, have about 50 tractors in their collection.

The old tractor was not without problems. By 1917 it had had two major engine failures. In 1917 a complete, new improved engine was installed at a cost of $900. It had not done a lot of work after that, so the engine was in remarkably good condition.

Irvin is continuing his research on the company and its history. The Pioneer 30-60, meanwhile, is handsomely preserved for posterity.

Old Tractors, Engines Help Him Let Off Steam

University of Missouri Vice-Chancellor Kee Groshong gets his jollies tinkering with antique Deeres and Case steamers

A vice chancellor at the University of Missouri, Kee Groshong is a John Deere aficionado.

There's no doubt about it, Kee Groshong loves old steam engines and gas tractors. In his collection he has a 1918 20-40 Case kerosene tractor restored, and he is in the process, with long-time friend Richard Adams, of restoring a 1917 65-horsepower Case steam engine. And, he has a half-scale model of the 65-horsepower Case.

When it comes to tractors, Kee is mostly a John Deere fan. He says he has "8 or 10" of the green breed with the oldest being a 1926 D on steel. The most recent Deere in his collection is a 1957

520; he also has a 1956 John Deere 420 crawler.

However, meet Kee at his professional desk and you'd never guess of his passion for old implements. Kee's impressive desk is in stately old Jesse Hall, the administration building of the University of Missouri in Columbia (MU). There, as vice-chancellor for administrative services, he is directly responsible for a $95 million budget. Kee is responsible for business operations; facilities operations and planning; academic and administrative computing; and operating MU's commercial television station, KOMU-TV, that also serves as a Missouri School of Journalism student laboratory.

But back to his tractors. Kee's interest in farm machinery comes naturally. He was born and raised in Lincoln County, Missouri, on a farm near Troy. There, Kee explains, his father and a vocational agriculture instructor kindled his appreciation of "mechanical things, things with moving parts, good workmanship, nice casting and foundry work, and superb

detailing work on old equipment."

"I grew up around two-cylinder John Deeres," Kee says. "During my youth my Dad had, at various times, an A, a B, and a 520. That explains my particular interest in John Deere tractors.

"And, I really like steam," he continues. "I clearly remember my Dad taking logs to Joe Schamma's steam-powered sawmill when I was 13 or 14. That was the first steam engine I had ever seen.

"I've never forgotten that experience. I was so impressed with that old steam

An officer with the University of Missouri's police force, Reese Groshong loves old tractors, too.

In their "working clothes" and a relaxed mood, the Groshongs get ready to pursue their favorite hobby.

engine. Joe would saw and blow the whistle, and it was really an eye opener for a young boy. I've been interested in steam since. As a matter of fact, I know where Schamma's steam engine, a 1910 15-horse Russell, is now . . . it's still in Lincoln County and it's shown at local events in my home community."

Jess Clonts, a veteran and widely respected vocational agriculture instructor at Troy High School, also had a significant influence, in many ways, during Kee's four years in vocational agriculture.

"Jess's influence had a great effect on me," Kee says. "He was multitalented. He had good judging teams. He took us alternately to the Missouri and Illinois state fairs. We went to the national and Missouri FFA conventions in Kansas City and at MU. Believe it or not, we also learned social graces under Jess's watchful eye.

"But, first and foremost, Jess Clonts was a shop man. He had a big shop. He had a mechanical bent and transferred that to his students."

With that background as a youth, Kee began his relationship with MU as a student majoring in accountancy and received a degree in 1964. He worked as

A vice-chancellor at the University of Missouri, Kee Groshong is a John Deere aficionado.

an accountant in a private business for a year and then returned to MU as a staff member in 1965.

His interest in old things was rekindled in the late 1960s when he attended the Missouri River Valley Steam Engine Show in Boonville, Missouri. Soon after, he got his half-scale Case steam engine.

As he became more involved in restoring old steam engines and tractors, he renewed a college acquaintance. While at MU, Kee says, he was roommates with Dennis Powers, now an Ogden, Iowa, farmer, John Deere dealer, and an avid "big time" collector of the old big tractors that broke the prairies, especially the Rumely, Case, and Aultman Taylor.

And, along the way, Kee got his son

Reese involved in his hobby. Reese, an MU police officer, is a partner in the Groshong restorations and the two, along with Dennis Powers, travel to antique tractor shows together. Reese says he started going to tractor shows as a young boy, six or seven years old. Over time, the enthusiasm and knowledge of the older collectors rubbed off on the young Groshong. He found that he had a natural affinity for things mechanical and he soon was involved working on the old machines. "I found I could make an old tractor run whether it wanted to or not," Reese says.

One challenge he enjoyed was working on a 30-60 S Rumely that was built to run on kerosene. "When it was restored by an earlier owner it was modified to use gasoline," Reese says. "I wanted to change it back to its original design—to run on kerosene. I had a real sense of satisfaction when I finished the job and had it running on the fuel for which it was designed."

Reese still enjoys going to shows and rubbing shoulders with the mostly older antique tractor buffs.

"There are three shows that we try to

attend on an annual basis," Kee says. "They are the Western Minnesota Steam Threshers Reunion, Rollag, Minnesota; the Threshermen and Collectors Show, Albert, Iowa; and the Missouri River Valley Show."

Kee takes an active part in the Boonville event. In 1993, Red Power was emphasized and it resulted in the largest showing of restored International tractors in the country.

"Lee Schmidt, an avid old tractor fan from California, Missouri, made a video of our 1993 Parade of Champions. Miles Wolfe, a Missouri collector from Pilot Grove, and I did the narration and more than a thousand tapes were sold by the end of the year," Kee says.

"The antique tractor hobby is one that has fascinated me," Kee says. "I enjoy being around the farmers, mechanics, machinists, and others that are attracted to the events. They're good people.

"And, if you're in a stressful job, you need some way to relieve that stress. I can leave the office and go home and read antique tractor and steam engine publications or work with Reese on a current restoration project and leave the office far behind.

"The hobby needn't be an expensive one," Kee continues. "You can buy an Farmall F-12 or John Deere A for around $500 and get in and have a good time.

"You can study history in many different ways and one is through the evolution of steam engines and tractors and how they transformed U.S. agriculture."

From Mail Box To Tool Box

Martha Stochl loves her postal job, but her day brightens when she gets back to her tractors

Hey, not just boys love those big farm toys! Martha Stochl, Toledo, Iowa, mail carrier, farmer, and dedicated collector of John Deere tractors, declares:

"From the time I was in grade school, I just couldn't get home fast enough after classes let out. I wanted to be around tractors and machinery.

"At Christmas, I wanted boys' toys, mostly. Farm toys. I remember my ranch set, too. I also loved horses, and until fairly recently, I bred and raised American Saddlebred horses. As my tractor involvement grew, I didn't have time for both. Now I'm down to one brood mare."

And the tractor numbers? "About 40—well, 40-plus. I haven't counted lately,"

Martha Stochl, of Toledo, Iowa, bought this John Deere 830 diesel—first one built—by sealed bid. She's pampered and polished it, keeps a factory shine on it.

The serial number on Martha's handsome 830 reads like this: 8300000. "Model 830 plus all zeros means it was the first one," she says.

she says with a laugh. As the numbers grew, she says, her parents asked, "What are you going to do with all those tractors?"

The collecting did squash a prediction made earlier by her brother. Martha says, "He told me that as soon as I started thinking about boys, I'd forget about tractors."

That didn't happen, obviously. Martha says, "I didn't have to think about my interest in machinery and mechanics. It just came naturally. I love to work in the fields. It makes me feel happy inside—the happiest feeling I have. My biggest dream would be to operate a bigger farm. Right now we farm 70 acres of the 240 we own, rent out the rest."

She prefers working behind the pop and chug of two-cylinder engines. Oh yes, she does use tractors from her collection for field work. Oh no, not her show models.

Martha takes four or five units to shows or meets. An 830 John Deere diesel tops her lineup. It probably came off the line in 1958. Martha says, "It's the first 830 built. The serial number plate reads 'Series 830, Serial No. 8300000.' "

She located it through an ad in a trade paper. The owner asked for sealed bids; she responded and forgot about it. She relates, "About a month later, I got a call, and the owner said, "You're the lucky one."

In fact, the 830 looks as if it came off the line yesterday. Martha, slender but strong, handles mechanical repair. She farms out body and paint work. Then she

Martha keeps her John Deere 435 diesel in showroom condition, too. Collectors covet this one. The 435 and 830 stop traffic at shows and meets.

concentrates on keeping tractors showroom sharp.

That explains why she can describe another Stochl cream puff, a 435 diesel (1959), as "better than new." Hand your mate the checkbook if you ever start looking at this one. But—nah, she'll never sell it. She did own three duplicates; sold them.

Her collection includes several unique models, including a 1938 AN, single front wheel. She also owns a 1959 730 standard diesel, with a gasoline "pony" starter engine.

Martha started collecting in 1983.

Surprisingly, that first purchase carried red, not green paint. She bought a 1954 Massey-Harris Mustang (original rear tires with Massey-Harris name). She still owns it, along with four or five other Masseys. But John Deere still ranks first with her.

Tractors aren't Martha's only hobby—she studied accordion for years and more recently bought a guitar and taught herself to play it; she loves to sing and dance.

But Martha will never be able to shake that collecting bug: "There are still tractors out there that I want," she says.

This Allis-Chalmers Was A First

A-C's first model might also be the first tractor

This much we know beyond the shadow of a doubt: The 1914 Allis-Chalmers 10-18 was that company's first entry into the tractor business. It had one forward speed and a reverse. It was a three-wheeler and its front wheel was off center. The body was painted green and its steel wheels came in red.

The particular 10-18 owned by Richard and Nancy Sleichter, of Riverside, in Johnson County, Iowa, was the first tractor ever in that eastern Iowa county, purchased by Joe Buchmayer from a mail-order catalog. A local historian, Irving Weber, records that Buchmayer made quite a splash

The 1914 Allis-Chalmers 10-18 was the first tractor model offered by this company. Owned by Richard and Nancy Sleichter, of Riverside, Iowa, this 10-18 was purchased in 1981 from the original owner, Mr. Joe Buchmayer, and is thought to be the first tractor in Johnson County, Iowa. The two-cylinder, horizontally opposed engine delivered 10 horsepower at the drawbar, 18 at the belt. It weighed 4,800 pounds and had one forward and one reverse speed of 2 ½ mph.

The Sleichters purchased this tractor from Mr. Buchmayer in 1981. "This was one of the biggest thrills of my life," Richard recalls. "I had always dreamed of owning a 10-18, and to be able to purchase one from the original owner was icing on the cake. Mr. Buchmayer was quite a guy. He was huge—6 feet, 10 inches tall, a very imposing personality. He was a Republican who got elected in a Democratic county. He was involved in organizing the 4-H, FFA, and the marching band. He was a visionary, and that, in part, explains why he was the first farmer in Johnson County to own a tractor."

Now, the question remains, is this tractor the very first one Allis-Chalmers rolled out of its plant in Milwaukee? "We really don't know. It might be but it can't be proven. We do know there are very few 10-18 tractors still in existence," Richard says. The tractor had been owned by Mr. Buchmayer for 67 years when the Sleichters got it, and it was in "decent" shape, although it had not been used for half a century or more. "I have a feeling Joe knew he should keep that tractor around and not let anyone disturb it or take any parts. It was easy to restore, everything was original, and the crank was still in the tool box," Richard smiles.

Both of the Sleichters are Allis-Chalmers boosters. Nancy's family owned and operated an Allis-Chalmers dealership for 50 years in Washington, Iowa. (The 1994 Classic Farm Tractors calendar, the one featuring their 1914

The strange front end design left the radiator isolated. The offset front wheel made steering difficult. Ads touted the tractor's one piece steel heat-treated frame—"no rivets to work loose."

Allis-Chalmers 10-18 for the month of May, was printed in Washington, Iowa. When A-C observed its 70th anniversary in 1984, their tractor appeared on the special commemorative calendar, along with other A-C classics.)

That same year, Richard went on a company tour with their tractor to promote the Allis-Chalmers name and products. "Also, the Ertl Company made a limited-edition models of the 10-18 for the Louisville Farm Show. It was quite a celebration, quite a year," Richard remembers.

Richard was in sales with Allis-Chalmers for several years, so his loyalty

when the tractor arrived by rail and he drove it to his farm, a couple of miles away. Wrote Weber: "It was indeed something of a novelty to the countryside, where all work was still being done with horses. Top speed of the tractor was 2 ½ miles per hour, but Joe was in no hurry. He was loving every minute of it." Mr. Buchmayer bought the tractor first; a car came later.

to the "Persian Orange" is unwavering. He has file after file of Allis-Chalmers information and literature, stuffed full like hogs in a finishing house.

He is a fount of information and is quick to recite some of the many accomplishments Allis-Chalmers racked up in its years in the farm tractor business. For example, A-C was the first to use rubber tires on tractors, and to promote the idea, professional race car drivers were hired to demonstrate its rubber-tire-equipped tractors at various state fairs. In June 1933 an Allis-Chalmers U with special high-speed gears, zipped around a Wisconsin track at 35 miles per hour.

In September 1933 world famous auto racer Barney Oldfield drove an air-tired Model U around a track in Dallas (a measured mile) at the amazing speed of 64.28 miles per hour. Thus, Oldfield, the first man to drive an automobile more than 60 miles an hour, accomplished the same feat on an Allis-Chalmers tractor.

Like all the other tractor makers that survived the critical early years, A-C made many changes, upgrading its tractors to meet customer needs. However, its first model, the 10-18, had a lot to offer from the starting gate.

A-C emphasized the no-rivet frame. It would not sag under any circumstances, and therefore motor bearings would never become misaligned because of frame weakness.

The 10-18 had a two-cylinder horizontally opposed engine, with two carburetors. The operator would start the tractor on gasoline and later switch it to a less costly fuel such as kerosene or distillate.

"The force-feed oiler was made by Detroit, the mag was a KW. The engine was cooled by a water pump driven by a long, endless belt. The belt begins at the flywheel, continues down through a hole in the frame of the tractor, up over a pulley that drives the fan belt, back to the flywheel again," Richard explains.

"When this tractor was introduced, it was equipped with foot brakes. Allis-Chalmers did not use foot brakes again on any models until 1938, when they introduced the WD.

"The pulley on this tractor extends beyond the left wheel. The purpose of this is that the motor is a left-handed motor, meaning that to run a corn shredder, corn sheller, or a small thresher machine with a twisted belt, you back the tractor up to the implement.

"Cost of the 10-18 was $2,200. Weight was 4,800 pounds," he says. About 2,700 units were built as Allis-Chalmer's entry into the tractor market.

The Sleichters usually take their 10-18 to two or three tractor shows in the area each summer. It is kept in the Mennonite Museum in Kalona, Iowa, the rest of the year.

At the Johnson County Fair in 1991, Richard noted a blue ribbon attached to a handsomely restored 1939 Allis-Chalmers Model B. The blue ribbon winner turned out to be Robbie Uthoff, a teenager, and the great-grandson of Joe Buchmayer.

If Joe is watching all this from above, you know he must be smiling.

This Man Is an Oliver Man

Jean Olson has shared his unflinching enthusiasm for Oliver tractors through tours, parades, and now an area museum

Most any day, two or three cars will stop at the Jean and Arlene Olson farm near Chatfield, Minnesota, to admire Jean's beautiful collection of Oliver tractors. They've had visitors from throughout the United States, Canada, and England, and from as far away as Finland, Australia, and New Zealand.

A local newspaper reporter once suggested that the "Olson farm set a county record for yield of tractors per acre." Jean admits to owning "around 100" tractors, but is quick to point out that not all are restored. A few are junkers that'll never see a fresh coat of paint; others provide parts. However, his spectacular collection of 80 restored Olivers sparkles like a new engagement ring, just as Jean's face sparkles every time he talks about Olivers.

His family was an Oliver family.

"My dad, Gus Olson, bought our first Oliver Hart-Parr 28-44 in 1930, the year they introduced the first vertical four-cylinder engine. This particular tractor had three forward speeds and burned either kerosene or gasoline. That was an excellent tractor and it really impressed me, so I got sold on the Oliver. Through the years, I bought 21 new Oliver tractors," he says.

Two of Jean's brothers worked in the Oliver factory in Charles City, Iowa, in the mid-1930s, and one later became a service representative for the Oliver Corporation in Minneapolis, traveling several states. Jean and the other brother eventually took over the family farm and at that point purchased an Oliver 70, one of Jean's all-time favorites.

Jean Olson loves to share his Olivers with others.

For many years his restored beauties have been a part of the Chatfield's Western Days parade, when as many as 30 of his Olivers have rolled down the parade route in the same celebration. At the August 1994 event, Jean was the parade's grand marshal, and his picture with his Oliver Hart-Parr 70 appeared on the buttons promoting the annual affair.

When the neighboring village of Fountain observed its 125th anniversary in the summer of 1994, Jean joined in in a big way. Having been on the county

It's very natural to see Jean Olson on an Oliver. He's bought, used, and collected Oliver tractors for many years—from 1930 through 1990.

This 1936 Oliver Hart-Parr 70 Row Crop was an exciting tractor in its day, and it still is. Owner Jean Olson, of Chatfield, Minnesota, has dozens of Olivers, and this is one of his favorites. Features included the first six-cylinder engine using either gasoline or kerosene, pressure lubrication, oil wash air cleaner, and a variable speed centrifugal governor. This was the last year Hart-Parr appeared with the Oliver name.

Valve-in-head engine operated at 1,500 rpm and had a 3 ⅛-inch bore, 4 ⅜-inch stroke. The 70 had four speeds forward, weighed 3,000 pounds, was 142 inches long.

board for 32 years and being on the board of the Fillmore County Museum, he decided to place three dozen of his tractors in the museum. Located in Fillmore, a community 30 miles south of Rochester, the museum is devoted to antique cars, farm implements, and all sorts of country memorabilia, once a part of everyday life in rural, southeast Minnesota.

When it comes to facts and figures about Oliver tractors, Jean spews out information like a computer. He can tick off the 15 "firsts" the innovative Oliver engineers introduced, such as the first to have a live PTO, the first kerosene-burning engine, the first tractor advertisement, the first tricycle row crop, the first valve-in-head engine, the first oil-cooled engine, and on and on and on.

Stroll with him through one of the buildings housing his collection, and he'll give you a play-by-play history of the Oliver company. "Hart-Parr was the first company to use the name 'tractor' . . . the Oliver Hart-Parr 70 was streamlined and modern- looking, setting a style for others to follow . . . the Oliver Super 99 was the most powerful tractor around in the early 1950s; it was equipped with either a diesel or gas engine."

Of his entire collection, the toughest tractor to track down was the Oliver 440, a 25-horsepower tractor and the smallest one he owns. When he was unable to locate one, he called the factory and learned that most of these models had been sold in the South. So Jean told his wife Arlene that they were going to take a little trip and drive until they found that particular tractor. Sure enough, Jean successfully captured his prize 440 in Kentucky. The round trip covered 3,000 miles—they saw a lot of the South—but most antiquers will confess that the hunt is half the fun!

Arlene has her own collection of Olivers, mostly toy tractors that their two sons played with as boys. "It used to be when you bought a new tractor, the dealer gave you a model of that tractor, too. That's how my collection got started. We also have six or eight pedal tractors, which have become very popular collectibles in recent years," she says.

The Olsons, who celebrated their 54th wedding anniversary in 1994, aren't as involved in collecting as in previous years, but they're always on the lookout for a bargain.

Through the years, the tractors Jean purchased arrived in various states of repair—good, bad, and indifferent. He farmed out much of the mechanical work to a former Oliver mechanic and hired a professional painter for the finishing touch. Of course, some of his

tractors still look and smell new. His "green model" of the White American series 60, for example, the newest one he owns came loaded with all the extras, including an air-conditioned cab. Quips Jean, "I drive it down the road on a hot day to tease the neighbors." It has never been used in the field, only in parades. Many of the antiques are displayed in an Oliver dealership setting, along one of the county roads leading to Chatfield. The collection is housed in several buildings. In all, there are some 300 tires to keep inflated—the steel-lugged and track models don't count. (His collection includes a smattering of Hart-Parr and Cletrac models as well.)

It may have been a calendar that sparked his interest in collecting. The year he started farming (with an Oliver 70), he received a promotional calendar picturing a farmer standing between an Oliver 60 and 70. He mentioned to Arlene that someday he'd like to own both those tractors. Obviously, his wish came true—and then some!

It was his 1936 Oliver Hart-Parr Row-Crop 70 that appeared on 1993 Classic Farm Tractors calendar, bringing Jean and his tractor full circle.

Young Farm Family Loves Old Iron

While Steve, Rachel, John, and Paul Rosenboom cater to Allis-Chalmers, their Friday takes the cake

Here's the Rosenboom family as they posed in 1983 for the Allis-Chalmers calendar and celebration of the company's 70th year in 1984. Rachel holds Paul, and John stands close to his father, Steve.

One of the nicest things about collecting and restoring antique tractors is that every member of the family can get involved. It's a tender trap that catches both young and old.

That's exactly the case with the Rosenbooms of Pomeroy, Iowa. Steve and Rachel have two sons, John, 15, and Paul, 11, and all four revel in the sport. Together they've searched for old tractors, reworked and renovated their prizes, and now all four attend shows, fairs, and parades sharing their enthusiasm—and unique antiques—with others, be they schoolchildren or great-grandparents.

You can't really categorize the Rosenbooms as an "orange paint" family, though the bulk of their beauties are Allis-Chalmers models. The first tractor Steve restored was an Allis-Chalmers WC that his grandfather had bought new and had been passed on to Steve's father when he took over the family farm. "I had no idea of the tedious process involved with restoration . . . the WC had been parked in a grove and was in terrible shape. But when I finally completed it and showed my dad the restored tractor that had been his father's, it was one of the proudest moments of my life," Steve says.

That sparked his interest, and his search for other vintage tractors began in earnest. "I've collected quite a few A-Cs, and an awful lot of friends," he smiles. "The thing I enjoy most about collecting old tractors is the people aspect. I just love getting around the people that grew up with this machinery at the various tractor shows. To me they are like walking history books."

The Rosenbooms' 1937 Allis-Chalmers WC with a factory wide front end and skeleton rear steel wheels always gets extra attention at shows because of its rare up-front look and because most people can relate to this particular tractor, which became so popular during its lifetime.

Then there's a 1937 orchard version of the M crawler (MO). The M crawler was a tremendously popular model for Allis-Chalmers in the late 1930s. Steve bought his from an almond farmer in central California, but as Steve is quick to point out, finding it was not the challenge—getting it back to Iowa was. The headaches were worth it, though, because it's such a unique model to the Midwest. "After all we'd gone through, we gave this particular tractor a professional paint job," he says.

When Allis-Chalmers celebrated its 70th year in 1984, the calendar commemorating the event pictured the crawler, as well as three other Rosenboom tractors, including a 1940 Model A. It's a stately looking tractor and was renown for its belt work.

A decade does make a difference; kids do grow up. This is the Rosenbooms in spring, 1994, with their rare 1948 Friday, an orchard tractor built in Michigan.

However, the tractor that gets the most attention at shows is a rare Friday, an orchard tractor built near Detroit.

Steve had seen a Friday at a sale in Michigan and later received a call from a man there who wanted to sell his 1948 Friday 048. Steve couldn't resist. Then he did some sleuthing and contacted the grandson of the company's founder, Phil Friday. The Friday Tractor Company is still in business, specializing in orchard equipment. It only built tractors for a short period after World War II, and every tractor was individually built and custom made. The Rosenbooms have found a second Friday; it's in the holding pattern, waiting to be restored.

The restored Friday is one of Paul's favorites. He and John were the "cover boys" on a *Successful Farming* cover in 1992, polishing the red tractor with the pug nose. Though he hasn't gotten to

This 1938 Allis-Chalmers WC is unique because of its wide front end and skeleton steel wheels on the rear. Owned by Steve Rosenboom, of Pomeroy, Iowa, he points out it was one of the most popular tractors ever, with production beginning in 1933 and continuing until 1948. It was the first to use a "square" engine, one having the same bore and stroke (4-inch). Its hand-operated brakes remained throughout production.

drive it in a parade (yet) he does get to wheel a 1956 WD 45 diesel down the street.

John and Paul both favor the big Albert City, Iowa, tractor show, partly because some of their school friends attend, and because they've met some new friends there as well. John's favorite tractor is a small G, the first one he remembers driving.

For Rachel, tractors have meant time together with Steve.

"I grew up with John Deeres and Farmalls, so when we first starting dating, the name 'Alice' kept coming up and I was beginning to get a little jealous. I soon learned it was Allis he was talking about. As a mechanic at the Allis-Chalmers dealership, he worked six days a week, so the time we spent together around tractors when he was off work was quality time.

"I like the way the boys interact with their dad around tractors. We've all gone to shows and camped out together where we can relax and just enjoy things. So, tractors are an important part of our lives," she relates.

Rachel and Steve both hold full-time jobs, she as a school secretary and he as an insurance inspector, besides farming 600 acres of corn and soybeans. "This isn't the way we envisioned life on the farm when we began farming in 1976, but we feel it is a great environment to raise a family," says Steve. They are the third generation living on the family farm.

Steve's mechanical prowess and past

experience with A-C works well for him in his hobby. He flat out knows tractors.

When he talks about the 1938 WC, he rattles off all the nitty-gritty details. "The WC first came out in 1933 and was produced through 1948 . . . unstyled until l939 . . . built in Milwaukee . . . the bread-and-butter tractor for A-C for many years . . . first in the industry to use a square engine: bore and stroke were each 4-inch . . . rated 14 horsepower on the drawbar, 23 on the belt . . . a four-speed transmission . . . it would go 18 miles per hour . . . it was smaller than other tractors but as strong because of its high strength steel . . . most other companies had gone to foot brakes, but it kept its hand turning brakes right to the end . . . the ads said it would pull two 14-inch moldboard plows in third gear at 4 ¾ miles per hour!"

One thing Steve doesn't plan to do is to race the Friday with any of his A-C tractors; he knows the Friday would win. "At shows" Steve says, "nobody ever believes me when I tell them it will

Climb on and take her for a spin. The Rosenbooms take the WC to several shows, where it's always a hit.

do close to 55 miles an hour, but I clocked it. That tractor will fly. It must have been the fastest tractor ever made," Steve believes.

Will John and Paul continue to be fascinated by old tractors? "Well," Rachel reveals, "they've already told me that if anything should ever happen to dad, don't sell the tractors!"

Their IH Mogul Is Part of the Family

Roger and Howard Schnell meticulously restored their father's one-cylinder tractor and it's always a part of their July 4 celebration

A 1917 International Harvester Mogul 10-20 is the centerpiece of the vintage tractor collection owned by the Schnell Brothers, Roger and Howard, of Franklin Grove, Illinois. "Dad purchased this tractor used at an auction sale in 1919," Roger explains.

It has been in the family ever since, and 70 years after their father brought it home from that sale, the two sons completely restored it.

From original sales material—some printed in color—and with literature from the IH archives, they were able to apply proper paint and lettering.

"There are only a handful of these tractors remaining, and after ours appeared on the Classic Farm Tractors calendar, we got calls from all over the country. Each caller had a 10-20 Mogul in various stages of restoration. A gentleman in Brochet, North Dakota, was restoring a Mogul and asked for the correct color code. We also received calls from Lynn, Washington; Middleton, Idaho; Fresno, California, and from Tilden, Nebraska. We even had two fellows call and request a photograph of the other side of our Mogul. Turns out

Howard (left) and Roger Schnell have a special affection for the Mogul they restored. It was purchased in 1919 by their father, Vernon, and remains in the family.

they were a woodcarver and a model maker and each wanted a picture so their renditions would be totally accurate," Roger says.

Whenever possible, the Schnell Brothers respond with pictures, information, and advice, sharing their expertise and experience—a commonplace,

everyday happening in tractor restoration ranks.

While some tractor collectors are wed to a single, specific make, the Schnell Brothers split their allegiance between International Harvester and Massey-Harris, again following in the footsteps of their father who died several years ago.

"Growing up, we had mostly IH tractors such as the F-12 and F-20" Roger and Howard relate. They are presently restoring an F-20 with a mechanical loader, a Farmall C with planter and cultivator, an IH No. 24 mounted picker, plus others "on the slow burner."

"Right after the war, in the late 1940s, our father really wanted a Farmall M—but so did a lot of other farmers. Dad was on the waiting list, but it was a very long list. There happened to be a Massey-Harris dealer in our town, too, and he convinced dad to buy his first Massey. Later, Dad added a Massey-Harris 44-6, then a Massey 55, and others followed.

"Roger and Howard have restored several Massey tractors, which they display each summer at the Franklin Grove Living History Antique Equip-

This 1917 IH Mogul 10-20 had a one-cylinder engine that operated on kerosene. There were some 9,000 of these tractors built from 1916-1920. McCormick and Deering had separate dealerships before merging, so there were two product lines. McCormick dealers sold the Mogul, and Deering dealers sold the Titan. The Schnell Brothers restored this tractor in 1989.

ment Show, a few miles from their Rising Sun Stock Farm. They have as many as eight tractors on display, but invariably, the Mogul gets the most attention with its primitive look and peculiar one-cylinder sound.

"Dad used it some for tillage and corn shelling, but most of its life was with a threshing machine. Though not the most powerful tractor, it was more than

A plate attached to the Mogul is revealing. It reads:
MOGUL ENGINE
USES
DISTILLATE, KEROSENE OR GASOLINE.
INTERNATIONAL-HARVESTER-CORPORATION.
TRACTOR WORKS - CHICAGO, U.S.A.
PATENTS PENDING
NO. BC5408 SPEED 400 H.P. 20

This tractor is interesting from any angle. It had two fuel tanks, gas for starting, kerosene for running. Hopper-type cooling system contains 35 gallons of water. The motor had an 8 ½-inch bore, and 12-inch stroke. It was equipped with a plow guide which aided in steering while plowing.

adequate for a small grain separator.

"It starts on gasoline, but burns kerosene. The fuel system has a water injection system which prevents pre-ignition or knocking. Its single-cylinder engine turns at 400 revolutions per minute and has an 8 ½-inch bore and 12-inch stroke. It has a hopper-type cooling system that contains 35 gallons of water; consequently, the operator is responsible for keeping it full. The water definitely evaporates under the load of the engine," Roger says.

The oiling system has four lines that feed the main bearings, connecting rod, and cylinder. The Mogul has a two-speed transmission. The tractor's speed is 1.8 miles per hour in low range, and 2.5 miles per hour in high.

"Dad was threshing in the 1920s, and one evening some

pranksters put a sign on the front of the Mogul, something about it being a slow, rusty old tractor. Dad retaliated with his own sign: 'Don't laugh fellows. Someday you'll be old, too!' As boys, we'd run the Mogul to celebrate the Fourth of July . . . we almost always drive it across the cattle yard and excite the cattle. While that wasn't such a good idea, getting the Mogul out for the family Fourth of July celebration became a fine tradition," Roger says.

The Mogul came equipped with a plow guide, which aided in steering while plowing. At the end of the field, it was raised so you could return on the next pass. It also had an over center actuated belt pulley clutch controlled by a small center cast wheel. The final reduction drive was done by an open chain that ran from the transmission area to the back where the differential was located. This was open and was a dirt-catcher.

Friendships flourish whenever and wherever people congregate to talk tractors.

In 1991, four members of the Ayrshire Vintage Tractor and Machinery Club of Kilmarnock, Scotland, visited the Schnell brothers. One, John Caldwell,

had attended a Massey-Harris show in Illinois a year earlier. They kept in touch, and in early 1994, Roger and his wife and another couple visited Scotland.

Roger was impressed by the quality of the tractors he saw there."About 15 to 20 percent of the collectibles are American, the rest being English, French, and German tractors. I saw several American-made models that came over during the Lend-Lease program during World War II. The height of prestige is a U.S. tractor well-restored and in good operating order. They are fine collectors, but they need a way to obtain parts for their American tractors. When our Scottish friends visited us, our agenda included junk yards. They were thrilled that old tractor parts were available," says Roger.

One of the highlights of Roger's trip to Scotland was a raffle held to raise funds for a tractor club. He and his friend had taken nearly three dozen seed corn caps, Franklin Grove buttons, pens, and the like. The items brought in about $180.

"We so enjoy the Scottish tractor collectors. They are a great group with high ideals and strong family values. That seems to be true of all the antique tractor people we meet, no matter the county, state, or country."

Roger concludes, "Tractor collecting is a very important part of our lives, and it all began with the Mogul. That tractor is just like a member of the family. If Dad is watching, he must be smiling."

This Family Runs Like A Deere

Bruce, Carolyn, Ashley, Casey, Brittany, and Jennifer Wilhelm all love John Deere tractors, regardless of the Deere's age or size

That old adage "opposites attract" wasn't the case for Bruce and Carolyn Wilhelm. At an early age, each was fascinated and thrilled by tractors, so this interest was a common denominator from the first time they laid eyes on each other.

Carolyn got to drive her dad's John Deere Model B when she was barely six years old (standing up, of course), and Bruce grew up with tractors, helping an uncle farm. So it's only natural their four daughters share this enthusiasm and passion for antique tractors, especially the John Deere kind.

The Wilhelm family, of Avondale, Pennsylvania, are such dyed-in-the-wool John Deere fans, that even the girls' names and initials reflect it.

The youngest of the Wilhelms, preschooler Jennifer Darlene, has the same initials as John Deere, and can't walk past a tractor without having to sit on it. She thinks all John Deeres are hers!

The Wilhelm family posed for this photograph with two of their John Deere tractors in the shop where much of the restoration takes place. Carolyn stands in the center behind Bruce who is holding Jennifer, 2. Ashley, 9, sits on a tractor tire, while the "drivers" are Brittany, 5, and Casey, 7. The entire family is intently interested in tractors—John Deere, that is.

At her show-and-tell sessions, Brittany shares tools used in restoring tractors. She takes bolts, clamps, and a J-hook used in towing. Brittany's tractor is a 720 diesel.

Casey hopes her parents will buy her a rare John Deere C sometime. They keep looking for one. Meanwhile, Casey has adopted an unstyled L—the L is for her middle name, Louise.

Ashley, the oldest daughter, writes reports about tractors, and though she's difficult to get up and off to school during weekdays, on Saturday mornings she's up and at 'em, accompanying her Dad to the shop to work on tractors. The day she got to help sandblast was a big day in her life. Ashley has a John Deere AW.

As a youth, Bruce, ran red tractors but longed to drive green ones. Why the attraction to Deere? "I think it's that certain two-cylinder sound. There's nothing else that comes close to it. I also like the idea that John Deere has never merged with another tractor company all these years," he says.

The antique-tractor bug bit Bruce one February day in 1984, while he was attending a farm sale near New Holland, Pennsylvania. Toward the end of a long row of tractors, he spotted a John Deere

One of the early specialty tractors built by Deere, the BO is right at home among that beautiful blooms of an apple orchard.

orchard model, "banged and beat up." Wilhelm was inspired by the fact that not many of this kind are restored, so he sold himself on the idea and bought the tractor.

When he got it home, he found the engine ran really well, so he pulled it and began the Herculean task of removing and sandblasting each component individually. It took him nearly four years to completely restore the 1936 John Deere BO, but Bruce admits "I only worked on it when I felt like it."

Nevertheless, this is one of the Wilhelms' special tractors because it appeared on the 1991 Classic Tractors calendar, selected by DuPont Agricultural Products from hundreds of entries.

Here's what Bruce Wilhelm says about his pride-and-joy John Deere BO.

"There were 5,054 of these tractors built, from September 1935 to January of 1947. Ours was the 434th tractor built; it has the brass serial tag on it.

"John Deere made the claim that this tractor could plow 9 acres in a 10-hour day. Today's tractors could probably plow that much in an hour. This tractor basically replaced four horses. It could pull two 14-bottom plows in normal conditions, so you could handle many jobs around the farm because of its size. It was easy getting on and off and had four forward speeds.

This 1936 John Deere BO ("O" is for orchard) is owned by Bruce Wilhelm, of Avondale, Pennsylvania, the 434th built. He completely disassembled the tractor, and each piece was painted individually. These models hugged the ground to protect against tree branches—even the driver's seat is lower. Water and fuel caps are shielded and muffler lays horizontally for protection.

"The high speed was 6 ¼ miles per hour, and reverse had a high speed of 3 miles-per-hour, which was pretty fast backing up. The horsepower on the belt is 18 ½ and at the drawbar she's 16 ¼.

"Its unique features include shielding up and over the fuel tanks, and a shield for the air intake, as well, to protect from tree limbs.

"Because these tractors were used in orchards, they were built a little bit lower than a normal tractor so it could get underneath the branches. The steering wheel and the seat are set lower. You could place the exhaust in one of any four positions.

"On this particular tractor, you started it on gasoline and then you switched it over to kerosene; then when you got done using it, you switched it back to gasoline so it was ready to start next time."

Bruce painted his BO with Imron because he feels it's the best finish and "it always has that wet look to it."

The second weekend in June is special for the Wilhelm family. That's when they host a two-day event for antique tractor enthusiasts, attracting some 200 exhibitors from the nearby states of Delaware and Maryland, plus half a dozen other states, some coming from as far away as Florida and Texas.

The Wilhelm Antique Farm Show starts with vintage tractors—60 or more of various sizes, makes, and shapes. In 1993, when they added a tractor pull, Carolyn was the only driver to pull the sled all the way up the hill, gunning Brit-

Contestants for the Kiddie Tractor Pull get final instructions.

antique autos (including Bruce's 1927 Pierce Arrow and Carolyn's 1955 Buick convertible) line the Wilhelm's front yard, near the main food stand where neighbors and relatives hustle hot dogs, hamburgers, finger food, lemonade, and all kinds of sweet stuff. It's like a big family reunion and picnic, with special sound effects.

tany's John Deere 720 diesel to victory. Special events such as pedal tractor races thrill kids and excite parents and grandparents like extra innings in a Little League baseball game.

There's also a great lineup of antique cars and trucks to admire, from an early Model T Ford pickup in prime condition, to a big, beautiful blue late-1940s Packard. Sometimes a man with a Stanley Steamer is there to give free rides to everyone in his surprisingly powerful, magnificent machine. The 30 to 35

Going like 60, Timothy Ammerman, 4, warms up for the big event.

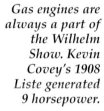

An injury (note Sesame Street bandaid) didn't stop Lauren Monarski, 5.

Gas engines are always a part of the Wilhelm Show. Kevin Covey's 1908 Liste generated 9 horsepower.

Cute kids—boys and girls—make any kiddie tractor pull fun for all.

Timothy looks like a winner whether he's on a green or a red tractor.

Sibling rivalry shows as sisters Lauren, 5, and Allison Monarski, 7, vie for the lead.

Casey Wilhelm, 7, on the Cub Cadet, "gets a lift" from David Foster, 10.

Wagon rides gave kids a lift throughout the day. Here Kenny Reed, 10, takes a turn.

Homemade and handy, this mode of transportation was one of many seen circling the grounds.

Just visiting and talking tractors is the main ingredient of any tractor show, regardless of size.

About 200 gasoline engines of every description sound off from their displays under nearby trees and on the exhibit grounds, much to the delight of veteran and novice show-goers.

Like hundreds of other antique tractor shows held throughout North America every summer, the Wilhelm affair is inter-generational, providing an ideal opportunity for the elders to share information and stories of bygone days with today's youngsters, in an atmosphere that's entertaining as well as educational.

The Wilhelms hold the show on a 5-acre piece of land adjacent to their service center where Bruce, Carolyn, and their employees repair cars and trucks, and where a heap of tractor restoration takes place after hours and on weekends.

For the Wilhelm's, a summer family vacation means attending a tractor show or parade or competing in a tractor pull, a sport Bruce took up in the summer of 1993. Winning his first couple of pulls was enough incentive to get him fired up about pulling, and it added a new dimension to their tractor collection and restoration activities.

All told, they have nearly 30 antique tractors, all but four of them John Deere. Besides the BO, their collection includes a 1940 B long hood, a 1938 L unstyled, and a 1945 BO Linderman.

At any given time, the Wilhelms may have three or four tractors in various stages of restoration, and these garner the interest of the entire family, including Carolyn's parents, Tom and Doris Jarrell, who restored the 1937 John Deere B that their daughter first drove. Bruce's father, Bob, though not a collector like Tom Jarrell, is a "cheerleader," and helped with the Classic Tractors calendar photo shoot when the BO was appropriately posed in an apple orchard in bloom.

Through membership in the national Two-Cylinder Club and their local Waterloo Boy organization, they met other John Deere fans and picked up leads on buying certain models they were seeking for their collection.

Their tractor trips have been something to write home about.

There was the time they struck out for

The adult tractor pull borrowed the sled belonging to the Rough & Tumble Club, of Kinzers, Pennsylvania.

Sporting his Wilhelm t-shirt and John Deere hat and suspenders, Bruce Powell, helped with the Two-Cylinder Club booth. Bruce is newsletter editor for the Southeast Pennsylvania Waterloo Boys.

South Carolina in a truck that had just been overhauled, but Bruce had not yet road tested. "That truck was ugly," Bruce remembers. "It was leaking fuel and the rear end was making so much noise we could hardly hear each other talking. But there was no turning back—we had to get to the sale on time! Carolyn was four months pregnant at the time. But we did buy our second tractor at that sale and got home safely, though we had our fingers crossed the entire 1,200 miles."

Returning from a sale in Indiana, his truck was overweight and the weigh station officer suggested Bruce "rearrange the load." Bruce says it's a mystery to him now how rearranging several tractors on a trailer could make

them any lighter, but sure enough they cleared the weigh station and headed home in a hurry.

Carolyn recalls the time she got a call at one o'clock in the morning from Bruce, who was on his way to a sale in Ohio. "Bruce had taken all kinds of truck parts just in case, but he had two blowouts on the Pennsylvania Turnpike and with only one spare he was stuck. I contacted the Turnpike dispatcher and told her my husband needed help, giving her a milepost Bruce remembered. Simultaneously, a trooper stopped to help Bruce and called into the same dispatcher, so the dispatcher had the trooper and me on the line at the same time. Nobody

could believe it," Carolyn says.

The trooper took Bruce to a truck stop, where Bruce bought a spare, then he returned Bruce to the truck, a very helpful gesture. Bruce made it to the Ohio sale (sleeping in the cab that night) and was able to buy the John Deere tractor he wanted, making the entire adventure well worthwhile.

Another trip that sticks in his mind was one in which he hauled several Fordsons to northern Pennsylvania for an uncle. To pay Bruce, the uncle gave him an old tractor that was in such terrible shape Bruce wasn't sure what it was.

It turned out to be a rare John Deere AOS, and the most memorable tractor trip he ever made.

This Swap Meet/Flea Market Has It All

This event is a must for serious tractor collectors: if you can't find it here, it probably never existed

It's a mecca for collectors of antique agricultural items. This event draws enthusiasts like clabbered milk draws flies to a hog trough. It's the Annual Waukee (Iowa) Swap Meet and Flea Market, billed as the largest agricultural heritage/antique fair in North America.

In 1994, in its 22nd year, the affair drew more than 15,000 people during its three-day run in the last week of May and included exhibitors from all 50

Framed by a tractor rim, Floyd Anderson (left) and Jeff Denekas have an assortment of tractor parts to sell or swap. Both are from Illinois.

Barb and Dave Kalsem are key players in the success of the Annual Waukee Swap Meet and Flea Market, located just west of Des Moines and held the last week of May.

states, Canada, England, Scotland, Australia, New Zealand, and Russia. There were more than 750 exhibitors in tight formation on the hillside facing Interstate 80 and behind the permanent building under shade trees, in tents, or on wide open spaces of this 47-acre tract

of land a stone's throw from Des Moines.

The antique tractor auction attracted some 80 tractors. "We had such a big response, we ran out of room," said Barbara Kalsem, president of the Central Hawkeye Gas Engine and Tractor Association, sponsor of the swap meet. Barb got lots of help from club members and her husband, Dave, who was the show director. Ordinarily, Barb Kalsem would be on the Iowa State University campus, where she is secretary for the department of agricultural engineering.

Barb and Dave note that this event has become so popular people begin arriving several days before the swap meet officially begins. Scads of people come in campers and recreational vehicles, and still the local motels are jammed. The place swarms with tractor collectors, restorers, and people looking for parts.

This gathering underlies the specialization that's cropped up in the past six to eight years as antique tractor collecting and restoring has become a white-hot sport—and business. There are now firms that rebuild and repair carburetors for old tractors; make gauges for antique tractors; manufacture tires for classic tractors.

If you want, you can place an order for a certain make of tractor, rough or restored. Would you like a tractor from Canada? That, too, can be arranged. Looking for some steel wheels or lugs? You'll find plenty.

The M. E. Miller Tire Company, of Wauseon, Ohio, is just who you need if you're looking for hard-to-find antique tractor tires and tubes. "We brought two large trucks to the show and half the tires were sold when we arrived," said Ed Miller. He explained that collectors call in their orders, then pick up their tires at the Waukee gathering, saving significant amounts of money by eliminating freight.

Orvis Wahl and Hans Knutsen, of Hutchinson, Minnesota, shared a truck ride to Waukee, and each returned with tires for their restoration projects. Orvis likes Deeres, Hans favors Case.

"We attend seven tractor shows during the summer in five different states," Miller says. Incidentally, his hometown is thought to have the oldest, continuous show in the country: The National Threshers Reunion in Wauseon, Ohio, marked its 50th anniversary in June 1994.

The Waukee swap meet is a natural site for business meetings of regional or national tractor clubs, since just about everybody's there. It's the perfect place to mix business with camaraderie.

At the antique tractor auction, it was wall-to-wall people. Some tractors were freshly painted, others needed a complete facelift, and as always the buyers wanted bargains. On this day in May, a John Deere High Crop fetched a final bid of $9,000; a John Deere orchard model brought $7,000; a John Deere L in great shape sold for $3,000. A Farmall High Crop went for $3,400; a Farmall M for $850.

Who knows what these tractors will bring next time around?

Just about anything you could possible want in the ag-related field is available at the Waukee Swap Meet and Flea Market.

The Waukee event attracts serious tractor collectors from all over the country. Some buy, some sell, some swap.

Jamie Stevens got a good look at the lineup of vintage tractors before the tractor auction got under way. About 80 tractors sold.

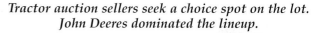

Foreign tractors always attract lots of attention. This Lanz Bulldog caught the eye of just about everybody.

Tractor auction sellers seek a choice spot on the lot. John Deeres dominated the lineup.

There was action galore during the antique tractor auction and the aisles were jammed with anxious bidders.

T-shirts make a statement—often declaring a favorite tractor. Justin Read, Richmond, Kansas, who sold the "Spoker D" t-shirts at Waukee, also shopped the grounds.

There are dealers at the Waukee Swap Meet from every state, and several foreign nations.

Call Him the "Diesel Man"

Like so many others, Alan Smith has had a crush on the Farmall M for years

The Farmall M received the highest number of votes when *Successful Farming* magazine polled its readers and asked them to name the "best tractor ever made." The survey was taken in 1984, three decades after the last M rolled off the factory floor.

Alan Smith, of McHenry, Illinois, is one of those admirers of the legendary Farmall M, but it was the diesel version that impressed him the most as a youth and continues to amaze him as an adult. "A neighbor bought this particular 1941 Farmall MD in the spring but couldn't take delivery of it until after the state fair

Alan Smith, of McHenry, Illinois, has worked with diesel tractors, trucks and earth-moving equipment for years.

in Milwaukee, where it was on display. The neighbor, Clint Raven, and his boys, did a lot of custom work for us, so I saw what this tractor could do. When I learned Clint had sold it back in the early 1980s, I sort of tracked it down. I wanted this tractor.

"When I got it home, we steam cleaned it and I completely disassembled the tractor. We completely rebuilt the tractor—new seals, bearings, sleeves, pistons, valves, and some used gears and

BEFORE: He'd had his eye on this tractor since youth. It needed some tender, loving care—and some paint!

AFTER:
This 1941 Farmall MD sparkles in the summer sun, just as the famous M tractor has sparkled through the years. A neighbor of Alan Smith, who did custom work, bought it new and Smith was impressed from the first time he laid eyes on this tractor. He was able to find and restore it later. International Harvester produced the Farmall M from 1939-1952, building about 288,000 of them, many of which are still in operation.

shafts. We sandblasted every single piece, replaced some of the sheet metal, and then we put on four coats of primer and painted it with Imron. It looked absolutely beautiful, and it still does," Alan says proudly.

The owner of a construction company, he's partial to the diesel engine. His heavy-duty construction and earth-moving equipment includes 10 Mack trucks and 27 Caterpillars, all diesel. His 10 antique tractors include two McCormick-Deering WD-40s, and John Deere 820 and 830 diesels.

Like thousands of others, Smith is enthralled with the basic Farmall M, diesel-powered or not.

What's the impetus behind this cultlike frenzy of Farmall M followers? Smith points to the powerful, smooth-running, throaty engine that performed, and just kept on performing. The tractor looked good; it was stylish, yet fully functional . . . like a fist in a velvet glove. It was a middleweight, weighing 3,500 pounds.

Even the price was right. A Farmall M with rubber tires carried a price tag of $1,112 in 1940; if you opted for all-steel wheels, the price was only $895. Electric starter and lights cost another $49.50, the Lift-All hydraulic system added an additional $35, and a swinging drawbar was $5 more.

The various regular and special attachments available for the Farmall M comprised a list a mile long.

International Harvester built almost 280,000 Farmall M tractors and another 8,300 high-clearance and specialty

models. It was a dominant tractor, a dominant model. Many are still working hard today. Alan uses his MD for mowing and raking and baling hay.

The Farmall mystique lives on. When DowElanco introduced a new corn herbicide, Broadstrike, in early 1994, it used the slogan, "It's a Whole New Era," and two-page advertisements in farm magazines showed the early Farmall M under this headline, "New Broadstrike. It does for weed control what the Farmall did for tractors."

When the Farmall M was introduced in the summer of 1939, it caused a sensa-

tion, along with its smaller mates, the Farmall H (two-plow) and the A (one-plow).

International Harvester had begun work on its new line several years earlier and hired Raymond Loewy to give these new tractors a new look. Loewy was of the Industrial Design school, sometimes

The diesel version of the Farmall M featured a 37 horsepower, overhead valve 4-cylinder engine with 248-cubic inch displacement and a 5-speed transmission. It was the first commercially successful row-crop diesel. An M cost about $1,112 in 1940. Electric starter and lights added another $49.50.

called "usable art." The concept called for taking utilitarian, everyday machines such as tractors, and making them more attractive, easier to maintain, and safer to operate.

Loewy gave the new red models a rakish appearance; he streamlined them by enclosing the fuel tank, steering, and radiator. He even had an orthopedic surgeon shape the tractor seat—pretty heady stuff for a rather stuffy industry. The overall shape suggested a combination of strength and agility, like a well-muscled athlete.

The M featured a 37-horsepower, overhead valve four-cylinder engine of 248 cubic-inch displacement, operating on gasoline or kerosene. Standard equipment included magneto ignition and steel wheels; however, electric starting and rubber tires were often offered as options, as was a new hydraulic system.

The diesel version of the M arrived in 1941, and the Farmall MD became the first commercially successful row-crop diesel.

Says Alan, "The tractor has a five-speed transmission, and in fifth it runs about 17 miles per hour. It was a three-plow tractor and was used for cultivation, mowing, disking, and everything else.

"It starts on gasoline, but it runs on diesel. You let the engine run for 20 to 30 seconds when the weather is warm or several minutes when the weather is cold. You then pull a compression release lever which raises the compression in the cylinders and advances the fuel injection pump; it runs as a full diesel engine," he explains. Easier starting was the whole idea. The right side of the engine had a carburetor and distributor, while the left side had an injection pump.

By 1952, the price of an M had more than doubled to $2,500 and the MD cost more than $3,200.

"My Farmall MD sold for about $1,800 when it was new. They didn't buy it with an electric starter, they crank-started it, so that's good enough for me," says Alan.

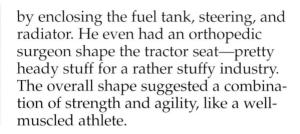

There's Beauty Beneath the Rust

A practical eye and a little faith allows mechanic-collector Lyle Spitznogle to find "gold" in old iron

A 1,000-mile "hunting" trip may reward collector Lyle Spitznogle with only a rusting hulk, but if he can make out the name "Hart-Parr," Lyle can see a gleaming show model in that corroded steel.

That keeps him searching and restoring when he can get away from his full-time mechanic's job at a White dealership. He lives near Wapello, Iowa. Although he does not actively farm, he keeps a small herd of beef cows.

As with many part-time collectors, he works within a limited hobby budget. He substitutes time, sweat, love, and his mechanic's skill to transform scrapyard candidates into show tractors.

He does well enough that his Hart-Parr 12-24 E graced the first page of the first edition of the widely famed DuPont calendar featuring photographs of classic antique tractors.

"That's my favorite," he concedes with a smile. "But I have a 13-30, kind of an oddball. When I get that fixed up, I might change my mind." He owns about a dozen tractors altogether.

Lyle got into collecting in the mid-1980s the way many have: "I started going to some of the shows, got interested, thought it was fun—grew to love it."

He collects only Hart-Parrs. Why? He explains, "Dad farmed with Olivers from about 1937. I used them at home. Hart-Parr had made tractors since about 1901 or 1902. They merged with Oliver in 1929; made the last Hart-Parrs in 1930.

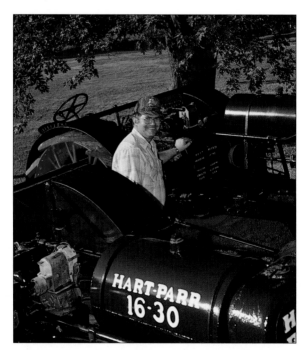

Lyle Spitznogle, Wapello, Iowa, would like one of every Hart-Parr model made, and says he'll keep at it as long as tractor collecting's fun.

"Actually, Hart-Parr stayed in the logo until 1937—Oliver Hart-Parr. So, [Hart-Parr tractors] are kind of unusual. Not that many around, but there are not that many Hart-Parr collectors."

Lyle likes the older models, built during and just prior to the 1920s. He admits to a fondness for the unique sound of the two-cylinder engines; a ka-chug, ka-chug different from that of a John Deere. Lyle says, "Their pistons fire at 90 degrees apart; John Deere's at 180 degrees."

When he restores a tractor, he does it all: mechanical work, sandblasting, painting. That increases his appreciation of the older tractors. "They're simpler, easier to work with; have greater tolerances. But it's surprising that builders did the job they did, considering the tools they used."

At the pace he must work, grabbing hours here and there, it takes him six months or longer to complete a restoration. Sometimes a job will wait on a parts search that might stretch into years.

He'd like to expand his collection to include one of every Hart-Parr model. "I probably never will get that done, but I'll keep trying," he says. His goal makes everything he finds a "keeper." Still, he will slick up a tractor for sale, occasionally. That gives him a little capital for more prospecting.

Lyle says family support has helped him to pursue his restoration hobby. "My wife, Kyla, goes to some of the shows with me. And she has been good about understanding my 'addiction' to old tractors.

"Our son, Eric, has helped me on a couple of tractors. And on his own, he has restored a pair of antique gasoline engines. Our daughter, Erin, lends a hand when needed."

Time will tell whether this attraction to old steel will carry into the next generation. Dad provides a role model: "I love the work. My greatest satisfaction comes from hearing a tractor run for the first time . . . taking a pile of rusty parts and turning it into a working, beautiful machine."

Below: They arrive a little rusty, but Lyle Spitznogle knows what it takes to transform a scrapyard ruin into a gleaming vintage tractor. He's done it before.

Above: Family support always gives a collector a boost. In this instance, Lyle's daughter, Erin, 14, lends a helping hand as dad works on a Hart-Parr 12-24.

It Only Takes One

*You need not have a huge collection to enjoy old tractor renovation:
Tennesseean Harold Glaus savors one F-30*

The folks at the International Harvester dealership in Murfreesboro, Tennessee, could hardly believe their eyes that day in 1942, when they saw an expectant mother checking out the F-30 that the Hoover family had traded in earlier for a Farmall M. She was at the crank and was obviously checking the F-30's compression.

The pregnant woman's husband returned in a few days and purchased

Harold Glaus, of Nashville, Tennessee, has a family tractor, and he knows its entire history.

the F-30, and not long after that, Harold Glaus entered the world. This tractor is Harold's pride and joy, truly a family tractor. His mother first located it, his father farmed with it for many years, and his son-in-law, Mike Flowers, helped Harold with the restoration. It's the only fully restored vintage tractor he has, proving that the fun and pride in saving old iron isn't measured by numbers.

Tracing a tractor's history is part of that fun.

"My Farmall Model F-30 was built in 1936, probably between Labor Day and Thanksgiving. The serial number is FB 17722. Not long ago I spoke with Ed Ryan, who worked the land with this tractor when it was new. He was employed by the Hoover family. Mr. Ryan remembered that it was on steel wheels and that the corn cultivators were difficult to put on. I assured him that I still had the steel wheels—and that the corn cultivators were as difficult to put on now as they were then," Harold grins.

"Dad bought the F-30 sometime before September 1942. He used the tractor one year on steel. In early 1944, he put Firestone rubber tires on the rear, and they're still on the tractor, have never had a flat, and still have the original calcium chlo-

ride in them," he says.

When Harold's father passed away in 1971, the F-30 was "semiretired." It was used sparingly for hay baling and disking. "A few times when other tractors in our neighborhood would not get the ox out of the ditch, I could always depend on the F-30 starting and completing the job that needed to be done," Harold muses.

In the late 1980s, Harold's son-in-law began the restoration after first doing some research. The tractor had been painted red, but was probably grey when new. Sure enough, some scraping exposed grey paint under the red. The tractor was in good condition and had been kept under cover. "It would be parked with the valve stems on the rear tires up, resting on timbers. This helped in making those tires see more than 50 years of life. The oil was changed every two days when it was working in the fields, and the engine had a complete overhaul in 1957. No body work was required; the tractor is almost completely original, including the four-bar steering wheel," Harold points out.

The McCormick-Deering F-30 tractors were built in Rock Island, Illinois. They were introduced in 1931, and production

Mike Flowers (left), Harold Glaus' son-in-law, had a strong role in the restoration of this 1936 Farmall F-30, the big brother of the F-12 and F-20 tractors. It would perform the tasks of a dozen horses back in 1931 when it was introduced to farmers, and it cost $1,075. The three-plow F-30 had a four-cylinder engine with a 4 ¼-inch bore and 5-inch stroke.

stopped in 1939, after some 28,000 were manufactured. Advertisements of the day stated the F-30 could do the work of 10 to 14 horses or mules, plowing up to 16 acres, planting 40 to 50 acres, or cultivating as much as 60 acres in a single day, all on only 25 to 30 gallons of kerosene. Versatile, an F-30 could also power a small sawmill.

An F-30 cost $1,075 new on steel. These tractors were built to last. Harold Glaus will attest to that, and so will hundreds of others who still have an F-30 around, helping with chores.

Recently, Harold and Mike acquired and restored a 1950 Cletrac crawler, a "fence-row tractor" abandoned in the brush for years, a real rust bucket. A 1957 Oliver Super 55 awaits major restoration.

Meanwhile, Harold, who runs a nursery business on the edge of Nashville, is happy to have the family tractor his mother, Clara, discovered at an IH dealership more than a half century ago, just before he was born.

Calendar and Computer Help Track Down 1911 Heider

Dick Collison and Omer Langenfeld are doing their darndest to find and restore three Iowa-made Heiders

The unexpected, long-distance telephone call to Dick Collison's home in Carroll, Iowa, that winter evening in 1991, came from Dan Eherding, another tractor buff in Ohio, whom Dick had never met. He had admired Dick's 1915 Heider Model C in the 1991 Classic Farm Tractors calendar and called to inquire about details of the restoration.

Near the end of their conversation, Dick happened to mention his quest for Heider Models A and B to complete the trio of tractors built by the Heider Manufacturing Company there in his hometown, beginning in 1911. (By 1916, the then-powerful Rock Island Plow Company had bought out the Heider tractor interests in Carroll.)

"I'm pretty sure there's a Heider A right there in Iowa," Dan said.

"Really? I've been scouring the country trying to locate one. Do you have any other information?" Dick inquired hopefully.

"Just a minute, let me check my computer."

Moments later he was back on the phone with Dick. Dan remembered reading an article in a magazine years earlier about an Iowan with a Heider A. Later Dan had entered the highlights of the story on his computer; he gave

It was a labor of love, as so many tractor restorations are, and it required 1,500 hours of work, but Dick Collison's hometown Heider was worth the effort.

Collison the owner's name, Harvey Soper, and the town, Anita, Iowa.

Dick was as thrilled as if he'd just won the lottery. Anita was only 55 miles south of Carroll. He hurriedly jotted down the name of the Heider owner from Mr. Eherding, and they said good night.

Next morning, Dick was on the phone. The operator informed him there was no one by the name of Harvey Soper in the Anita phone directory. Disappointed but undaunted, he called the newspaper publisher in Anita, who was stumped by the name.

He did, however, give Dick the name and phone number of an octogenarian, Meta Miller, a lifelong native of the community, who "knows everybody in the county."

Dick finally got through to her. "I don't know of any Harvey Soper, but there is a Dave Soper, and my guess is that's the man you're looking for," she said. The trail was getting warmer.

"Do you have an old Heider tractor?" Dick Collison inquired of Dave Soper when at last he got him on the line.

"Yes, I do," Mr. Soper replied, "and I plan to restore it myself."

"Could I see it?," Dick asked.

"Why sure. When?"

"How about in the morning?"

This 1915 Heider Model C was built in Carroll, Iowa, and some of the Heider family children still living there provided information about the tractor. Its most unique feature was its variable-speed friction drive. It came with a Waukesha four-cylinder engine with a 4 ½-inch bore and 6 ¾-inch stroke, developing 10 horse-power on the drawbar, 15 on the belt. It was 12-feet long, weighed 5,500 pounds, and cost $995.

Collison and his partner Omer Langenfeld were at Soper's place bright and early, and when they walked into Mr. Soper's shop, there hung the Classic Farm Tractors calendar. Thus, Collison knew that Soper knew he had restored the Heider Model C, and was genuinely interested in the Heider they were about to see.

But Mr. Soper wasn't ready to part with it; he still harbored thoughts of restoring the 1911 Heider Model A.

To say the Model A needed lots of work is a gross understatement. The Heider had been abandoned in a field under a tree for 60 years of its life and was, as they say in the trade, a rust bucket. However, Collison and Langenfeld, had patents, manuals, literature and advertisements about all three Heider models built in Carroll because two sons of the manufacturer still live there.

Together they fit the pieces of the puzzle together, using the 82-year-old

Two years and several visits to Mr. Soper's farm passed, and on one such visit Dave said that he realized he didn't have time to restore the tractor himself, and he was willing to let go of it, knowing it would have a good home in Carroll, where it originated.

It would step right out and do 4 miles per hour, though its normal speed while plowing was 2 miles per hour. This 1915 Heider C is displayed at the farm museum sponsored by the Carroll County (Iowa) Historical Society.

promotional pieces as blueprints for the tractor parts they had to build. Work was completed in the fall of 1993. "We think [the Heider Company] only made 25 to 40 of this first model, so it may be the only one left anywhere," Collison says.

The handsome, mint Heider Model C was not as tough a chore for Collison and Langenfeld. Gaylon Anderson, a farmer near Marathon, Iowa, had owned it, and another C was donated by Earl Buske, of Humbolt, Iowa, to be used as a parts tractor.

The Models C's unique feature was an infinitely variable friction drive from the engine flywheel. No gear shift. No clutch. No transmission.

"Speed was controlled by levers in the cab which slid against the friction drive disk. When the speed was changed, the radiator, engine, and hood all would slide. Move it toward the driver to go faster, away to go slower. It would go a 'blazing' 4 miles per hour, but its normal speed was 2 miles per hour when plowing and pulling three-bottoms," Dick says.

That beautiful green machine with red and yellow trim is displayed by the Carroll County Historical Society at a museum/farmstead. Its four-cylinder Waukesha engine with a 4 ½-inch bore and 6 ¾-inch stroke generated 10 horsepower on the drawbar, 15 on the belt. The Heider Model C was 12 feet long, with a 96-inch wheelbase. It weighed 5,500 pounds and listed for $995 if you picked it up in Carroll.

Collison and Langenfeld spent 1,500 hours restoring the C, strictly a labor of love.

"Our hope is that we will have all three models of the Heider tractors that were built in this town for our grandchildren and their grandchildren to know and enjoy," Dick says.

The missing model is the B, and the search goes on to find one to bring back to the little town where it was built. They will not rest well until that third tractor is added to the restored A and C, making the Iowa trio of Heiders complete.

A retired veterinarian and businessman, Dick Collison has found doctoring old Heider tractors much to his liking.

This 1911 Heider Model A was sought out and restored by Omer Langenfeld (left) and Dick Collison—after their prized 1915 Heider was on display. Now, they hope to find and restore a Heider B, last of a trio of tractors built in Carroll, Iowa, early in this century.

Boredom Led to Love of Old Iron

Missourian Zachary Lee Schmidt is one of the most enthusiastic (and knowledgeable) tractor collectors in the country

I f Lee Schmidt hadn't been bored on a hot Friday afternoon in August 1984, he might not now be the avid antique tractor collector he is.

The boredom came while the Schmidt family, of California, Missouri, was visiting at the farm home of Raymond and Pearline Stephens, parents of Lee's wife Ruby, near Bush City, Kansas.

To take his mind off of the Kansas heat and to relieve his boredom, Lee went for

Lee Schmidt is from California—California, Missouri, that is. He favors International tractors, but some of his best restorations have been of other makes.

a walk around the Stephens farmstead. He passed a line of old farm machinery in a fence row that he had seen many times before on previous visits. This time a rusty tractor caught his attention. He noticed it was a John Deere; he wondered if it would ever run again.

That was the beginning.

When he returned from his walk, Lee told his father-in-law about his idea to get the John Deere running again. He explained that he could work on the restoration project whenever he and Ruby made one of their frequent visits to the Stephens' home.

"Dad told me to have at it," Lee says, "but warned me that the Deere had been in that fence row 14 years.

"Dad also told me it was a 1937 John Deere B and that he and his father had bought it new in 1938 for $410 and a horse."

By asking questions Lee found out that the B originally had skeleton

steel wheels, which were later replaced with rubber tires. The steel wheels had been sold as scrap iron to help the World War II war effort. The tractor had been parked since 1970, when the magneto bracket broke.

With his father-in-law's encouragement, Lee went back to the John Deere for a second look. It was in bad shape. The engine was stuck. The radiator had a fist-sized hole in it. Water pipes and hoses were rusted and rotted. The hood had rust holes. The front tires were flat and the rims were rusted through. The muffler and exhaust pipes were missing.

"But, I decided then and there that the tractor would run again," Lee says.

"We pulled the tractor to a spot under a shade tree and I was in my element—a shade-tree mechanic. I went to work tearing it down to unstick the engine.

"The next day my brother-in-law, Wayne Anderson of Topeka, Kansas, came by. Wayne fancies old tractors, especially the IHC variety, and he was glad to see I had gotten interested. Wayne and I worked two days on the Deere, and with the help of a sledge hammer, a 4- x 4-inch block of wood, and a lot of perseverance, we broke the pistons loose.

This 1930 Allis-Chalmers U is one of Lee Schmidt's proudest possessions. It was selected for the Classic Farm Tractors Calendar in 1991. "I love old iron, and I'd walk a mile just to look at an old tractor." The U sold for about $900 new, depending on options. It had a Continental S-10 engine with 284-cubic inches. Allis-Chalmers also produced the 1930 U for the United Farm Equipment Company.

"Even though we had made real progress I was discouraged because of all the parts needed and distance from home. But," says Lee, "Wayne wasn't about to let me quit. He encouraged me and sent me a subscription to *Gas Engine Magazine*.

"We were back in Kansas in September and Dad surprised me by telling me to take the John Deere home with me so that I could really work on it without interruption," Lee continues. "So, I hauled it back to California.

"By July 1985, almost a year after starting the project, I had the tractor restored. I thought it was beautiful and named it 'Pearline' after my mother-in-law," Lee says. "A month later I was ready for my first tractor show. Wayne invited me to bring Pearline to the Heart of America Steam Engine Show at McLouth, Kansas.

"Even though I was apprehensive, we had a really fine time and it turned out that Pearline was one of the nicest tractors in the show.

"In September I took Pearline to an antique tractor show at Cole Camp, Missouri, and she won the trophy for Best Restored Tractor on Rubber in a field of 18 entries.

"I was so encouraged that the very next day I went to the Missouri River Valley Steam Engine Association show at Boonville, Missouri, and was amazed and pleased when, from a field of nearly 150, Pearline won the trophy for Best Restored Tractor.

Here's a rare one—a 1948 Farmall H-15 cotton picker, serial number 4540. Schmidt plans to exhibit it at the Red Power Roundup in 1995.

This Farmall H will shine like a new silver dollar when Lee Schmidt gets through with it. It will, undoubtedly, be worth several silver dollars at that point.

"By now the bug had really bitten me and I was reading everything I could get my hands on. I talked tractor with anyone who would listen or talk. I was ready for a second project, and it was a 1936 IHC Farmall F-12. Even though it was in bad shape, I had it ready for the 1986 show season.

"And, by 1986 I had found a set of skeleton steel wheels for Pearline with the help of my new tractor friends and *Gas Engine Magazine*. I took her back to Bush City, Kansas, where it had been purchased, for that city's centennial in June 1986.

"Dad drove that 1937 John Deere B in the Bush City Centennial parade," Lee says. "It was the first time he had driven her in 16 years."

By now Schmidt was firmly hooked on "old iron." He started looking for another project that he could afford, and that led him to an early 1930s John Deere "Junior" hay press. He completely restored it so that it baled beautifully, and it made an excellent companion piece for Pearline. Lee showed the pair at the 1987 Boonville show.

His next project was another John Deere B that he bought from a friend in 1988. It was missing two wheels, tires, and other parts, and the motor was locked up. During the engine overhaul Lee found he had something special: the engine was a nonproduction, experimental prototype made by Deere in 1939. No parts were ever available for it, and, as a result, Lee's friend, Hubert Koenig, Boonville, Missouri, made replacement valves out of D-6 CAT valves by turning them down on his lathe.

By the time Lee was finished, the B had factory spoke wheels, tires typical of the era, fenders, a front wheel mud scraper, umbrella bracket and umbrella, plowing clevis, tools, PTO drive tire pump, and other special Deere accessories. He also restored a John Deere plow, a two bottom 12-inch plow on steel wheels that he pulled behind the B at special events like the Expo III in Waterloo, Iowa.

Lee's real prize, however, was yet to come. He had heard of an old tractor from a friend and got in touch with the owner to see whether he wanted to sell it. The owner, Pat Kline, of Sedalia, Missouri, invited Lee to visit and look at the tractor, which was at an abandoned farmstead. Parked in the fence row since 1955, the tractor was by now well camouflaged: A cedar tree 11 inches in diameter and 15 feet tall had grown up through the machine between the fender, firewall, and transmission and through the steering wheel. The steel wheels had settled until the oil pan was resting on the ground.

Kline told Lee the tractor was an Allis-Chalmers and that his father had purchased it new in the early 1930s. The owner wouldn't price the tractor and turned down Lee's offer of $500. Early in 1989, however, Lee was able to trade Kline a New Holland 56 hay rake that he had completely restored for the Allis.

After he had the A-C home, Lee determined that it was a 1930 Model U with a Continental engine; its serial number was 4824. He also learned that only 7,404 of the tractors with that engine were built between 1929 and 1932.

Lee's home show, the Missouri River Valley Steam Engine Show, was featuring A-C tractors for its 1989 event, and he was determined to have the U fully restored in time for it. With help from friends and the former owner, who kept finding missing pieces, Lee made the deadline with a few days to spare.

Since Lee was in the habit of naming his restorations, it didn't take him long to decide that the U was a jewel, and so he named it after his jewel of a wife, Ruby.

Ruby—the tractor, that is—was a smash hit at the Missouri Valley show.

"As I was unloading her I had a cash offer that only a fool like me would refuse," Lee says. Ruby was the winner in her class and went ahead to win first place in four other shows in the fall of 1989.

However, the highlight, as far as Lee is concerned, came when John Harvey, originator of the DuPont Classic Tractors calendar, called from Wilmington, Delaware, to ask whether he could use Ruby as the March pinup for DuPont's 1991 calendar.

In recent years Lee has become an International Harvester collector, not only of numerous tractors but other products made by that company as well. He has an International freezer, refrigerator, snow blower, signs, and other pieces in his collection.

His current restoration project is an International one-row cotton picker that he bought at Bragg City, in Missouri's cotton-growing country, the Bootheel. The picker is mounted on a Farmall H, which was built to run backward. Lee's goal is to have the picker, already in fairly good shape, completely restored so that he can take it to the 1995 Red Power Roundup to be held in Des Moines.

Lee is a member of the International Harvester Collectors (National and Worldwide) and is founder and current president of the IH Collectors, Missouri Chapter. The Missouri chapter, along with the Missouri River Valley Steam Engine Association, were cohosts of the fourth annual 1993 International Harvester Red Power Roundup. Lee was chairman of the event; the Roundup drew the largest collection of International Harvester items ever assembled in one location.

On his business card Lee confirms that he is "a lover of old iron" and that he is "dedicated to preserving part of our history for the future." The card also reads, "Tractors—all makes and models—look, talk, buy, sell, trade, collect, and restore."

"I'm hooked on restoration as a hobby, and as a business, too," Lee says. "I'll continue as my income permits, and with two children in college, that has been a factor."

However, Lee has the support of his family. His son Chris has helped on restorations. Also, while wife Ruby may not share the same enthusiasm for an old tractor that Lee generates, she travels to many of the shows and thoroughly enjoys the people at tractor events.

Lee firmly believes that his hobby has helped him develop as an individual. "I'm not the same person I was 10 years ago. Then, I dreaded getting up in front of a group of people and speaking. Now I enjoy it.

"I've developed numerous lasting friendships with collectors all over the world. Tractor people are some of the greatest on earth. Most are honest, helpful, caring, and they make it all worthwhile."

Dick Bockwoldt:
Tractor Restorer Par Excellence

By turning out some of the most admired tractor restorations in the country, this Iowan has them standing in line for his touch.

Unknowingly, Dick Bockwoldt's doctor made him happy by telling him he'd have to stop farming because of his bad back. Bockwoldt took his physician's advice: he stopped farming (almost) and began restoring antique farm tractors on a nearly full-time basis. And he's enjoying every moment of it.

Should you want him to resurrect your old tractor, you'll probably be put on a waiting list. Because of his meticulous craftsmanship, attention to detail, and caring attitude, tractor buffs beat a path to his door.

Dick and Dorothy Bockwoldt's farm, not far from Interstate 80 in eastern Iowa, lives up to its name, Antique Tractor Acres.

Approaching their place from the south, a driver spots long lines of tractors—some like retirees seated in a doctor's office awaiting their appointments, others like bread loaves on a bakery shelf ready for purchase.

Several of the barns behind the Bockwoldt's neat farmhouse are crammed with unique antiques, some restored,

some not. The large pole-frame building that houses Dick's repair shop and paint room has 8 to 10 tractors on hold outside, and that many or more inside in various stages of restoration.

It's a hustling, bustling atmosphere.

Recently he was working on four aged John Deere tractors at once, all of them GP (general purpose) models, the official tractor of Expo IV, sponsored by the Two-Cylinder Club, in Grundy Center, Iowa. Each GP was barebones, awaiting its turn to be repaired, refurbished, and repainted. The GPs belonged to four different collectors in four different states.

Dick owns several fully restored 1928 to 1935 John Deere GP tractors himself, including one with a matching three-row planter, another with a three-row front cultivator. The third was one of the last GP built. While Dick is somewhat partial to Deere, the farm equipment he's always used, it's his habit to track down various rare makes and breeds. These include Huber, Rock Island, General Ordinance (GO), and Happy Farmer. If the truth be known, Dick

Dick Bockwoldt is all smiles over the toy tractor collection he has amassed, including a John Deere he played with as a boy. Though he has many Deeres—large and small—Dick collects rare tractors, too.

Bockwoldt has never met a tractor he didn't like, and before he's finished he'll probably restore at least one of just about every tractor listed in the history books.

His business is booming. A Midwest restaurant chain, the Machine Shed, calls

Rock Island tractors are special to Bock-woldt because he grew up 20 miles from the factory where they were made in Rock Island, Illinois. His 1929 Rock Island FA has a four-cylinder Buda engine. He has collected eight Rock Island tractors, and admits he's "always looking."

and orders "a green tractor and a red tractor" to decorate the entrance of its new eatery in Rockford, Illinois. An acquaintance pulls in unannounced with his father's tractor, unloads it from his truck and tells Dick, "This was Dad's first tractor and I want you to restore it. No hurry. Just call me when it's finished." Another call comes in from a tractor collector who wants Bockwoldt to restore a rare model he's found. Eventually he plans to show it at the Mt. Pleasant, Iowa, Old Thresher's Reunion, and he insists that Bockwoldt do the restoration. (More and more of the show-room ready, Bockwoldt-restored tractors appear at major shows.) Another fellow calls to announce he's bringing a truck-load to Dick for restoring.

"It does get hectic at times," the affable Bockwoldt admits. "When I have a deadline to meet and can't find the parts, that's when it gets a little hairy around here. My daughter, Denise, reminds me that once in a while we are putting decals on the night before the show, but that doesn't happen often. There's always plenty to do, though, that's for sure."

There are some people in tractor circles who question the deep, shiny finishes Bockwoldt insists on; some call it the "wet look" and suggest the old tractors never looked that good.

Bockwoldt is adamant about the finishes he applies. "To begin with, cheap paint fades and discolors, so why go to all the time, expense, and work to restore a tractor, and then have a less-

Huber tractors are another favorite of Dick and Dorothy Bockwoldt. Their special jackets say so emphatically.

than-perfect paint job?" Dick is also quick to point out, "These finishes are so much easier to keep clean, and when you use a hardener, you don't have stains."

He has perfected a paint system that involves some rather expensive products, and some surfaces get as many as 10 coats.

Dick does most of the engine and body repair work, relying only on specialists to handle block and head repairs and magneto and carburetor restorations. He admits that when he tackles a tractor project, "I like to do it myself so I can control the result."

Of the nearly 100 tractors on his place, the rarest is a 1920 GO, built by the General Ordinance Company at Cedar Rapids, Iowa, which took over the Denning Tractor Company when the latter went bankrupt. It has a four-cylinder engine and cost $1,400 new. It weighs just a tad over two tons.

The Bockwoldt's married daughter, Denise Hedberg, helps put the finishing touches on restorations by carefully applying decals after Dick's completed the painting.

Another of his favorites is a 1918 Happy Farmer, built by the La Crosse Tractor Company, of La Crosse, Wisconsin. It has a single wheel up front, a two-cylinder engine, and a sporty orange color. When first introduced in 1916, the Happy Farmer sold for just $695. Later models were priced at $1000 to $1,200.

Dick likes to collect Rock Island tractors, partly because they were manufactured in Rock Island, Illinois, only about 20 miles from where he grew up. He owns eight. His 1929 Rock Island FA is a show-stopper, with its handsome gray body, red steel wheels, and yellow lettering. It sports a four-cylinder Buda engine, rates 18 horsepower on the drawbar, 35 on the belt. (The Rock Island Plow Company was purchased by the J. I. Case Company in 1937.) Soon after Dick restored them, he displayed this tractor with a plow at a historical event in Moline, Illinois (which neighbors Rock Island) and at the Farm Progress Show.

Not only does he enjoy restoring tractors, he loves finding them. Anytime Dick's traveling, he's looking, hoping to spot an old tractor behind a barn or showing through from inside an open door of an outbuilding.

He says one of the first tools collectors sometimes use is a chain saw. "Lots of old tractors are discovered that have been abandoned in a grove for years, and trees have grown up, around, and through them. Getting them loose and then getting them loaded can be another adventure," he smiles, "usually not recommended for bad backs."

Once in a while, however, he gets lucky. "I sort of stumbled onto a rare 1918 crossmotor Huber in a horse barn in Ohio. Of course, digging it out was no fun—those horses had left their calling cards all over the place."

Bockwoldt attends several major tractors shows annually, including the Midwest Old Threshers, Mt. Pleasant, Iowa.

A non-coffee drinker, he once went to pick up a Rock Island tractor, and the owner insisted Dick have a cup of coffee before he left. "I did sip it, but I haven't had a cup since and don't plan on having another one," he claims.

Unless, of course, there's a tractor with it.

Show Horses Versus Work Horses

The Alexander brothers have different ideas about what they want from restored tractors

Mark and Russ Alexander have a lot in common. The brothers jointly manage a sprawling farm operation in Ralls and Marion counties, in northeast Missouri. They both have a keen fondness for old iron with green paint. They both are members of the Vintage Green Association, a regional affiliate of the Two-Cylinder Club Worldwide. And they both own well-maintained tractors that now are pushing 60 years of age.

Here, however, the brothers' interests begin to diverge. Mark owns a 1935 John Deere B that looks as if it just rolled out of the factory. He wouldn't think of asking the vintage tractor to do even light chores around the farm.

"The only nonstock item is an extra fuel shut-off petcock at the carburetor," he says. "Otherwise, it's completely original. I've shown it at local and state fairs, and it's a regular feature at parades."

Russ, on the other hand, has a 1937 Deere A with an engine that has been bored and stroked to deliver nearly four times its original horsepower output.

"I like to hear an old tractor hard at work," he says. "I pull the A in 4,500- and 5,500-pound classes all over the country, but I wouldn't take a tractor that was an ideal candidate for restoring and turn it into a pulling tractor. At least, I don't think I would."

Despite their differing philosophies about what old tractors should be and do, the Alexander brothers support each other in their particular passions. Russ proudly boasts about the plaques and trophies Mark's B has won as the Best on Rubber at antique tractor shows. Mark is

Mark Alexander (on tractor) and his brother Russ share a mutual passion for vintage John Deere tractors, but have widely differing motives for rebuilding them.

one-man rooting section when his older brother hooks the modified A onto a pulling sled.

"This tractor was built on March 19, 1935," says Mark of his meticulously restored B. "It's a brass-tag tractor, serial 8,600."

For some reason that has been misplaced in history, the company affixed brass identification tags to random tractors: Mark's B wears one of those brass tags. Deere began serializing the B with number 1,000, making Mark's the 7,600th B built.

Mark has tried to trace the history of his tractor. From the factory, the machine was delivered to a dealer in St. Louis. Ownership eventually fell to a farmer near Mexico, Missouri, whose son was interested in restoring it to original condition.

"Then, the son was killed and the father lost interest," says Mark. "But he kept the tractor in excellent shape. A few years ago, he sold it to me."

Mark points out unique features on his B.

"It has rubber tires on spoke wheels,

front and rear," he says. "It was one of the first B's to come out of the factory equipped with rubber tires. Most of the older ones on rubber now are on cut-offs—wheels that were originally steel, then had the steel rims cut off and tire rims welded on."

The original-equipment Firestone 9.5 x 36 tires on Mark's tractor still have the molding tips.

"The tractor still has the original tool box, mounted on the frame just in front of the engine," adds Mark. "And I found a practically new plow-clevis drawbar."

Mark even located original John Deere silk screens when he painted the bright yellow signature on his tractor.

Russ Alexander admires his brother's dedication to authenticity, but Russ had a different task in mind when he began modifying his John Deere A.

"I bored the cylinder to 5 ⅜ inches from 5 ½ inches and used pistons from a Model 70 Deere," he says. "I built a new crankshaft, to stretch the original 6 ½-inch stroke to 8 ⅜ inches. I had to cut away part of the transmission case so the connecting rods would clear."

Russ also put in a higher-lift camshaft and a G carburetor, along with a different head and bigger intake and exhaust valves.

"I've spent more than $3,600 just on the engine," he says. "But I've got it doing what I want it to do now."

What it does is turn more than 80 horsepower on a dynamometer—quite a boost from the rated 23 horsepower the Model A originally boasted. When Russ

"I don't expect my Model B to do any work," says Mark Alexander. He enjoys owning a 60-year-old tractor that is virtually the same as it came from the factory.

Mark's 1935 John Deere B was an early model to wear rubber.

Cranking up for another tractor pull, Russ spins the flywheel on his hot rod Model A.

Russ Alexander beefed up his 1937 John Deere A so it turns out more than 80 horsepower, compared to its initial 23. "These tractors were built to work hard and that's what I expect them to do," Russ says.

juices his old tractor with a special racing fuel mixture at antique tractor pulls, he usually roars into the money.

Isn't he worried the souped-up engine will put too much strain on the A's running gears?

"Naw. The drive trains on these old tractors were really overbuilt," he says. "Even with more than three times the original engine power, there's not much danger of damaging the transmission or differential. These tractors were built to work hard, and that's what I expect mine to do."

Mark Alexander nods in understanding, but he wouldn't consider overboring the engine on his restored B. Not for one minute.

"I guess I'm more of a purist," he says. "My tractor is for shows and parades. I just like knowing that I've got a 1935 model tractor that is virtually the same as it was when it came off the assembly line. It doesn't have to do any work."

He's the Dean of Tractor Collectors

Lyle and Helen Dumont's dream museum mixes tractors with Trigger

Lyle Dumont has pursued and restored antique tractors for almost 30 years. "I've traveled everywhere in the U.S. and up into Canada . . .. I'll go anywhere to buy one."

As his interest grew, so did a dream. His wife, Helen, shared it: someday, maybe, a museum, a place to show off the cream of Lyle's growing collection.

Now, talk with any serious collector and the word museum will tiptoe into the conversation; the Dumonts, however, weren't content with talk. And so it was, with faith and much hard work that the Dumonts were able to open Dumont's Museum of Dreamworld Collectibles.

Located near Sigourney, Iowa, the initial steel structures surround a whopping 24,000 square feet, and that number will probably grow. "I can see the

Unique design of this 1952 Oliver 77 orchard tractor makes it one of the favorites of Helen and Lyle Dumont, of Sigourney, Iowa—the "apple of their eye", you might say. The Dumonts have a large collection of pain-stakingly restored tractors, emphasizing Olivers.

museum expanding to 80,000 square feet in 10 years," Lyle says. "We expect it to grow into our retirement occupation." Who said you have to take it easy when you retire?

Antique tractors headline the "show," centered on Lyle's specialty, Olivers from the 1940s, 1950s, and 1960s. Carefully refinished, they gleam with the diamond-bright luster that marks a Dumont restoration. Lyle says, "We aim to make a tractor look better than it did out of the factory."

Why Olivers? His dad farmed with Oliver and White tractors. One of those Olivers serves as Lyle's chore tractor today. He likes them. He adds, "Not much interest when I started collecting them. But they did well in antique tractor pulls, and that gave them attention. They're hot with collectors now." He auctioned off 65 of them in the fall of 1993—and started looking again!

He began collecting tractors as a teenager, but he says interest has gradually shifted forward. Now many collectors want tractors they can parade—and that calls for rubber tires.

The first 100 tractors into the 60- x 200-foot building also included Lyle's oldest, a 1915 Case, two-cylinder opposed. Also on exhibit are a Waterloo Boy, several Rumely OilPull, and Hart-Parr models, and others.

"We'll boost tractor numbers to 200. And we'll invite other collectors to bring in tractors on a rotating basis. Same with antique cars," Lyle says.

There's another fascinating aspect to

Lyle Dumont (right) believes restoration should make an old tractor look better than it was brand new. He has restored more than 500 tractors. Here, he checks the paint job on an IH piece with employee Dave Finch.

their museum. The Dumonts allotted a wing for their assemblage of Roy Rogers and Dale Evans commercial memorabilia. Lyle says, "We think it may be the biggest collection of this material outside the Rogers' museum in Victorville, California."

They know them well and have spent many hours visiting with Roy and Dale, Lyle says. "We go out there twice a year," he adds. "We usually stay two to three days, and spend several hours at the museum. They're wonderful people, the best. Religious, and the most down-to-earth folks you'd want to meet."

The Dumonts also have made space for a large collection of John Wayne memorabilia, purchased more recently. Lyle hopes that their museum, as it develops, will take visitors from the present through the recent past and—with the western movie connections—help them relate to earlier times.

Lyle and Helen have structured the museum as a nonprofit organization. Lyle hopes for wide community involvement. He says, "We want something nice here, something that will entertain and educate, and bring people into our community."

With the museum a flourishing reality, Lyle can reflect on how it all started. He worked in an Oliver dealership while he attended high school. Later, he repaired tractors at an IH dealership. He began to collect, and restored the tractors as a hobby.

He acquired sandblasting equipment to help strip paint and clean metal. That ability led to other jobs and full-time employment as a sandblasting and painting contractor.

Lyle and his employees honed their skills at restoration and painting all the while. His own antiques exhibited a gloss that attracted custom business. He

This 1940 Oliver 80 Row-Crop Diesel is one of only 75 built, and only four are known to exist. This was the first year Oliver produced a diesel tractor. Lyle Dumont is the third owner, having purchased this rare tractor in Arizona nine years ago. The diesel engine was a Buda-Lanova combustion system with American-Bosch injection. Later, Oliver replaced it with its own diesel engine.

weeks. A lot of sheet metal? Three weeks or more. We have to do it on a time-and-material basis."

Helen stays close to the restoration business, although she holds a full-time job as a nursing home administrator. "I used to do more," she says. "I painted all the wheels, and did the lettering."

She operates an Oliver decal business that Lyle founded. He says, "There weren't any [decals available]. So I had them reproduced, and we sell them all over the world."

Helen notes, "We get orders from South Africa, Australia, New Zealand, lots from Canada, and quite a few from England. We get tour groups from England." Count on tour group numbers booming in years ahead, as the museum grows and word gets around.

As the museum takes shape, Lyle continues all aspects of his business. You know he won't drop search and restoration activities anytime soon when you consider his reasons for doing them:

"I find a real challenge in locating and buying old tractors. You may see a rusty old hulk, but I can picture it when it's done, and see how nice it's going to look. That's what gets you excited. I don't see the rust, really.

"Quality guides me, especially for this museum. I want top restoration, top paint. I want quality above quantity."

Not a bad guide for any business.

says that he spends about 25 percent of his time on custom restoration, 25 percent restoring units he owns, and the remaining 50 percent on his "full-time" occupation.

The numbers add up. Lyle says that he has restored more than 500 tractors. The time investment depends on the tractor, he says. "If a model has little sheet metal, we might turn it out in two

He Keeps an Eagle Eye on Case

Dave Erb, editor of Old Abe's News, devotes countless hours to the J. I. Case Collectors' Association (JICCA)

Santa Claus himself would envy this guy's beautiful beard. Mr. Claus might also be envious of some of Dave Erb's credentials. Erb is a writer, editor, and publisher, producing *Old Abe's News* on a quarterly basis for Case tractor buffs everywhere. He's also a teacher, a mechanic, and a former service manager.

David Erb is coauthor of the book, *Full Steam Ahead—J. I. Case Tractors & Equipment 1842-1955 (Volume One)*, published by the American Society of Agricultural Engineers. This book is indicative of his love for the Case company, its early development of farm machinery and steam engines, and its later manufacture of numerous gas tractor models, many of which are classics.

A history lover, Erb and his wife, Sandra, live in a log home that was built in the 1800s and is listed in the National Register of Historic Places.

His magazine is "for folks who like J. I. Case history, people, and equipment," and each issue lives up to its statement of purpose. A typical edition will be 40 or more pages with a color cover featuring a handsome Case tractor. Inside are letters from Case fans, various feature stories from Case archives, and reports on past

That's David Erb (behind the beard) with one of his students, Donald Pearce, who helped tear down, put back together, and paint this 1929 Case L. Erb is editor of "Old Abe's News," a publication for Case collectors in North America and elsewhere.

and upcoming Case events, shows, and reunions.

Because Case tractors were painted different colors through the years, a Case paint color information guide helps restorers determine correct colors for their tractors—from Flambeau red, desert sunset, or LC grey to any of the other colors used for Case tractor bodies and wheels.

There are JICCA members throughout North America, as well as Australia, New Zealand, Iceland, Denmark, Sweden, Germany, Holland, England, and South Africa.

In 1989, Erb acquired a 1929 Case L from northeastern Kentucky. The tractor had come from an Indian reservation in Wisconsin and had found its way to the Mid-South.

The tractor became a shop project for Erb's students at Buckeye Hills Career Center, Rio Grande, Ohio. It was in rather sorry condition, and the students laughed about its prospect for restoration.

To begin with, the engine was frozen. However, students watched and learned as it was broken loose with a wooden block and hammer.

Two senior boys, Don Pearce and Ray Bragg, took the lead and tore into it. "Don nearly jumped clear over the workbench when the unmufflered engine finally barked to life," Erb remembers.

In the Ag Mech shop the boys took it completely apart, sandblasted every

The 1929 Case L featured the first in-line Case engine. Because of its smooth running engine, it was extremely popular with sawmill operators and threshermen. The new four-cylinder overhead valve motor with a 4 ⅝-inch bore and 6-inch stroke developed 26 horsepower at the drawbar. Case tractors were built in Racine, Wisconsin.

piece and did all the things a pro would do to prepare it for painting. Then they used an epoxy primer, regular primer, and four coats of finish—all under the watchful eye of their teacher.

The Case was transformed from a sad sack to a smashing success. So good was the restoration job, it qualified for the 1993 Classic Farm Tractors calendar.

The 1929 Case L claimed several firsts. It featured the first in-line engine, as the company switched from the crossmotor design. In its first year, more than 6,800 units were made. It had a three-speed transmission and used a unique design that involved roller chains in the differential drive. An oil bath air cleaner was another feature. The four-cylinder overhead valve engine started on gasoline but ran on kerosene, which at that time was much cheaper than gas. It developed 26 horsepower on the drawbar. Smooth-running, it was loved by sawmill operators and threshmen.

"When first introduced, the Case L and other tractors of that era, came equipped with steel wheels. But when its production ceased in 1939, practically all tractors came from the factory with rubber tires," Erb notes.

Dave is often asked why the magazine is called *Old Abe's News*, and what is the significance of that term? Dave has this explanation:

"During the Civil War, patriotic feelings burned hot in the hearts of Americans on both sides of the Mason-Dixon line. Visiting Eau Claire, Wisconsin, in 1861, a 42-year-old businessman from

The newly-designed engine started on gasoline but ran on kerosene, which was the "inexpensive brand."

Racine, Wisconsin, by the name of Jerome I. Case, witnessed the mustering of Company C of the 8th Wisconsin regiment.

"This company's mascot was a massive, live bald eagle, named Old Abe in honor of then-President Abraham Lincoln. Fierce in appearance and defiant in nature, this bird developed the amazing ability to assume the correct demeanor for any military situation. In parades he was especially spectacular, responding to martial band music with a flurry of wings, sitting impressively on his perch, a shield of the United States union colors.

"This mascot went on to become famous, participating in no less than 42 battles. His presence served to inspire soldiers of his unit. As his fame grew, his presence at battle lines so infuriated the Confederate armies that rewards were allegedly offered for his death.

"After the war, Old Abe became a sought-after guest at patriotic and political gatherings and invariably caused more excitement among attending crowds than did many a guest speaker or visiting dignitary.

"So impressed was Mr. Case by this fierce mascot that after the war, in 1865, Old Abe became the official trademark for the J. I. Case Threshing Machine Company.

"Old Abe eventually became the trademark for our country as well, serving as the model for the U.S. federal bird, the Marine Corps, and several other armed forces groups.

"Old Abe still graces equipment of the J. I. Case Company. He serves as a timeless symbol of integrity and honor, for a company and a country. This publication is proud to share that tradition, dedicated to the proposition of serving Case collectors the world over," Erb concludes.

He Collects Olivers, Large and Small

Dale Johansen designed his house with his hobby in mind

He retired from farming in 1988, but Dale Johansen will never retire nor relinquish his quest for Oliver tractors, real or miniature.

He and his wife, Lois, live in town now but are only a few minutes' drive from the Latimer, Iowa, farm where he has three quintessential Oliver tractors, a 1945 Oliver 70 Standard, a 1946 Oliver 60 Standard, and a 1936 Oliver Hart-Parr Row Crop 70 with tip-toe steel on the rear and rubber wheels up front.

"This Row Crop 70 was like the first tractor my father owned, and I earned my wings on it. I've driven Olivers ever since, and at one time owned 24 pieces of Oliver equipment, including a combine, corn picker, sheller, and manure spreader—along with an assortment of plows, discs, cultivators, and rakes.

"When dad bought the Row Crop 70, it had steel wheels in front, too, but in turning at the end of a row, those wheels would dig up corn plants something awful, so he replaced the steel with rubber," the Iowan says.

In another nearby building, there's a 1952 Oliver 77, the tractor Johansen started farming with when he returned after a stint in the armed services. He'll restore this one for sure—in time.

A long-time toy collector, Dale Johansen is strictly an Oliver buff. The display room in his Iowa home showcases his collection beautifully.

Another pair of Olivers, a 1947 Oliver Row Crop 70 and a 1948 Oliver Row Crop 60, also await restoration. Dale has several pieces of Oliver equipment that he plans to paint and spruce up to hitch to these tractors when they're restored.

"I like the idea of showing restored, antique tractors with equipment of the same vintage. It gives people a more complete picture of how things were done."

A 1966 Oliver 770 waits nearby, its

Oliver front-end loader ready for the day's chores.

Just mention Dale's collection of toy Olivers, and his face beams. Visit his specially designed display room and you feel as though you've entered the F.A.O.

A complete set of Oliver tractor models, plus numerous other pieces of memorabilia, stock his display room.

Johansen, of Latimer, Iowa, has been an exhibitor at the National Farm Toy Show every year since its beginning.

Schwarz store on New York's Fifth Avenue. The 35- x 15 ½-foot room is that impressive.

One entire wall is devoted to shelves of model tractors and miniature pieces of farm equipment. A mirrored wall behind the collection doubles the effect. Four display cases hold an array of Oliver collectibles. One of these is a prized oval, pocketbook mirror from the early part of the century; the opposite side of the mirror pictures James Oliver with his name and company, "Oliver Chilled Plow Works, South Bend, Indiana U.S.A."

There are 1,200 model tractors, 800 implements and trucks, and 150 cast-iron toys. A trade magazine once referred to him as the "Toy Man." "I actually built this house for my collection," Dale says. "It was the only way I could get a new home," Lois chimes in.

Adjacent to the handsome hobby display room is Dale's workshop, where he's always working on another project. He has made a number of Oliver tractor models of wood, as well as wagons, machinery, and a surrey. His shop is complete.

"I never go halfway," he smiles. "I have all the boxes my Oliver tractors came in, too. Sometimes they become as valuable as the tractors."

He has seven Oliver pedal tractors in perfect condition. An assortment of Oliver signs and

This handsome Oliver 70 is one of several restored models Dale owns, and several other tractors await restoration. He also has numerous pieces of Oliver farm equipment and machinery.

Lois and Dale Johansen are Oliver fans all the way. Besides farm tractors, they collect toy tractors, and exhibit at the National Farm Toy Show each year.

calendars grace one wall. There are belt buckles, watch fobs, and other memorabilia. That's not all. He has a complete collection of advertising pieces and Oliver tractor and machinery literature. One ad shows an early Oliver gang plow—54 bottoms stretching across 64 feet in one gigantic furrow.

"You name it and I've got it if it's Oliver. Not many know Oliver made outboard motors for boats, but here's the proof," he says, pointing to a model and accompanying promotional piece.

"I started collecting seriously in 1955, and really got into it after meeting Bob Gray, of Eldora, Iowa, one of the pioneers in building toy tractor models. At that time, things were still affordable. Everything is so expensive now, but that's to be expected as more people get interested and the demand increases. Pedal tractors, for example, can bring several thousand dollars each. It's a little crazy, but I love it," he says with a wry smile.

At a recent show he found and purchased a large model truck, exactly like the one his father had brought home to him 65 years earlier when his dad took a load of cattle to market in Chicago. The truck, however, has not made it to his "Oliver Room"; it isn't an Oliver!

World-Class Farm Toy Town

Toy tractors and the name Ertl give Dyersville, Iowa, worldwide recognition

In the critically acclaimed movie Field of Dreams, filmed in a corn field just outside of Dyersville, Iowa, there's a brief but memorable exchange between two characters:

"Is this heaven?" the first asks. "No, this is Iowa," comes the reply.

To the thousands of people who love and collect toy tractors, Dyersville itself must seem like heaven. It is the home of three premiere toy makers: Ertl Toy Company, the world's largest manufacturer of farm toys; Scale Models, headed by Joe Ertl; and Spec Cast, owned by Dave and Ken Bell.

The National Farm Toy Museum, with more than 25,000 pieces, is located in Dyersville, and the Summer Toy Festival (June) and National Farm Toy Show (November) are held here, swelling the town's usual population of 4,000 by four or five times during "show time." It's almost as though the founder of the Ertl enterprise, Fred Ertl, Sr., had responded way back in 1945 to another famous line from Field of Dreams, "If you build it, they will come."

And come they do to the toy shows, to tour the toy plants and to buy toys at several retail shops and outlet stores. It's an immense, international business. Ertl has plants in Mexico, China, and Italy, a procurement office in Hong Kong, and sales offices in Canada and Great Britain.

The Ertl Company, a subsidiary of Hanson Industries, occupies almost 500,000 square feet of office and factory space and employs nearly 1,000 people in Dyersville. Though much smaller in scope, both Spec Cast and Scale Models are expanding their operations as well.

For tractor model and farm toy enthusiasts, Dyersville, Iowa is a "must see." The National Farm Toy Show is the only place to be in early November. That's the case with Iowan Dale Johansen who has been an exhibitor every year since the event began in 1978. A dyed-in-the-wool Oliver collector (see page 98), Dale not only sells a number of items, but also finds several goodies at this giant toy extravaganza in Dyersville.

The National Farm Toy Show is held in November and it is usually winter gloves and parka weather.

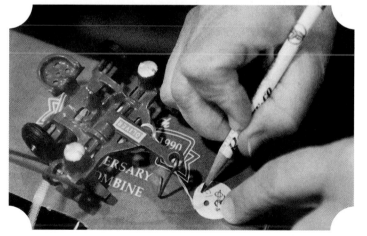

They never miss—Dale and Lois Johansen have been faithful exhibitors at every national show.

Above:
Johansen's booth specializes in Olivers.

Below:
Special prices make buying irresistible.

As many as 20,000 toy tractor enthusiasts fill an entire gymnasium and spill over into several other buildings.

The end-of-the-show auction offers unique and special models for the serious collectors. Some who attend the auction are buyers at every auction.

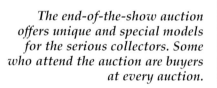

Every item gets close scrutiny from potential buyers.

A Farmer and Former Farmer Enjoy Tractors

Bill Dreisbach, Jr., Pennsylvania farmer,
and Roy Ashley, landscape architect in Atlanta, both love tractors

Take a tour of the Bill Dreisbach's home and it's almost like taking a tour in a John Deere museum. There are Deeres everywhere: toy model tractors stretch across the fireplace mantel from end to end; lovely pieces of John Deere artwork hang on the wall (he and his Model D appear in one); in a living room chair there's a pillow with embroidery of a Johnny Popper; downstairs there's still more. "This is a John Deere house," Bill says, smiling broadly.

Dreisbach farms some 400 acres in southeastern Pennsylvania, using John Deere tractors and equipment almost exclusively.

Meanwhile, in Atlanta, Roy O. Ashley, of Roy Ashley & Associates, commercial landscape architects and land planners, fondly recalls his days on the farm and the Allis-Chalmers tractors he grew up with. He now collects A-C tractors and equipment. He has a good start on a collection of the combines built by Allis-Chalmers. In fact, he has one of every pull-type model that was ever made by A-C between 1937 and 1962.

And like most farm boys who grew up on a farm, he's never forgotten the tractors. "Those unique pieces of iron

He's never seen a John Deere he didn't like, but this 1935 Model D is his favorite. Bill Dreisbach farms in southeast Pennsylvania, and collects plows as well as tractors (he has 15 Deeres).

were so much a part of the lives of farm people. At the time, we hated, kicked, mistreated, cursed, and wore out many of them, but deep in our hearts, they

were loved so much that they really became 'one of the family,'" Roy remembers.

Both men have a passion for their

It's a John Deere kind of Christmas at the Bill Dreisbach farm with five green Deeres pulling Santa's sleigh. Rudolph is a riding mower, then there's an unstyled L, a GP, a 1935 D and finally a 1947 D. Note antlers on the Deeres.

particular tractor "breed." Dreisbach is a John Deere fan, while Ashley adores the orange paint of Allis-Chalmers. Though they've never met, if they ever should, they'll no doubt talk tractors till the cows come home.

Bill's favorite tractor is his 1935 John Deere D. Roy's favorite is a 1937 Allis-Chalmers UC.

Bill says his D has always been easy to start, no matter what. He's owned it a dozen years and when he bought it, "that D was solid rust. There were no fenders and the body was in terrible shape. However, the bearings and rings

These striking "before" and "after" photos from the family album emphasize the amazing things that can happen when a tractor is renovated. This is Bill Dreisbach's junker turned parade tractor.

were fine, and mechanically the tractor was in decent condition."

The before-and-after pictures of Bill's tractor attest to the fact that collectors can transform old, ugly equipment into gleaming restorations. He drives the tractor in parades, takes it to local shows, and just flat-out enjoys it.

Bill points out that the Model D was the first to carry the John Deere name and it was made for 30 years, longer than any other Deere.

Roy Ashley purchased his Allis-Chalmers UC in Lovejoy, Georgia. "The tractor was completely junked, and by its appearance, it had already gone to the graveyard. I suspect it was

Roy Ashley is an Allis-Chalmers supporter, who's collecting A-C tractors, equipment, and combines. Roy, a landscape architect with a firm in Atlanta, remembers his rural roots.

purchased by the Farmers Home Administration and placed on a farm in Georgia as a demonstration of new technology and efficiency when compared to the horse and mule."

Roy points out that the UC was the first row-crop tractor produced by Allis-Chalmers, though many think the WC holds that distinction. It was known for its ease in implement changing; it had a drive-in cultivator, a new power take-off system, and other features. "The significance of this particular tractor is that it was 'all tractor,' with immaculate simplicity, lasting power, and no frills. The motor was bolted directly to the transmission and become one unit. Hand brakes were standard and it ran on low-cost fuel." Roy says.

The Allis-Chalmers UC rated 12 horsepower on the drawbar and 25 on the

The Allis-Chalmers UC was that company's first row-crop model, and its "drive-in cultivator" beat the competition by a country mile.

Strictly functional, there were no frills on this tractor. The motor was bolted directly to the transmission and it became one unit. It rated 12 horsepower on the drawbar and used low-cost fuel.

pulley. Production of this model began in the early 1930s and ended in 1951.

Meanwhile, back in Pennsylvania, Bill Dreisbach is also pursuing the antiquity and importance of the plow. He owns a wooden plow that dates to 1763, probably from Berkshire in England. His collection includes horse- and oxen-drawn plows that helped pioneers tame the great U. S. prairie, and he takes these to shows with his John Deere tractors.

Two men with differing professions and varying tractor preferences, but with a common hobby. Each is helping to preserve an important part of our agricultural heritage.

Vintage Tractors Make a Great Fair Better

At the Missouri State Fair, mint-condition antique tractors have created new excitement

It all began with a brainstorming session. Dianne Larkin, publicity director for the Missouri State Fair, was searching for ways to expand and upgrade the extravaganza with Larry Harper, editor of the *Missouri*

Ruralist, the Show-Me State's major farm magazine.

Harper had an idea. "Hold an antique tractor show, emphasizing completely restored entries. Call it the 'Best of the Best,'" the editor suggested.

Larkin immediately liked the idea. Harper offered his magazine to publicize the event and provide trophies for the winners. What they needed was someone to spearhead the project.

Both remembered Gordon Chrisen-

Kingpin in the "Best of the Best" antique and classic tractor show at the Missouri State Fair is Gordon Chrisenberry, an avid International collector.

Dianne Larkin, Director of Publicity for the Missouri State Fair, learned to drive a tractor before she drove a car. She's involved with the tractor show.

Editor of the "Missouri Ruralist," Larry Harper, himself a tractor fancier and collector of agricultural antiques, illustrates "how you find `em" in the weeds.

Harper and Chrisenberry are different. Both were born and raised on farms in western Missouri and were introduced to tractors early on.

Chrisenberry grew up in Vernon County near Nevada, Missouri, where his dad worked for an International Harvester implement dealer.

"In 1939, when I started high school and enrolled in vocational agriculture, my dad let me farm 20 acres," Chrisenberry said. "I bought a 1930 Farmall Regular, plow, disc, cultivator, and three Jersey heifers for $750.

"I still have the Regular. I've restored it and people tell me it doesn't look like a Regular. But, it looks the way it did when I bought it in 1939."

Chrisenberry's farming operation was interrupted by military service in World War II, and during the Korean War his reserve unit was so sure it would be called to active duty that Chrisenberry sold his farm in preparation.

The call never came. So he found a smaller farm and combined farming and over-the-road truck driving to earn a living. Now semiretired, he has developed a new occupation. During many spring, summer, and fall weekends he is kept busy announcing events at fairs, tractor shows and pulls, parades, and other similar events around the state, something both he and Mrs. Chrisenberry enjoy.

And, he also works on tractor restorations.

His specialty is the International line, and in his restored collection are Interna-

berry, an active member of the Chilhowee Antique Farm Machinery Collectors in central Missouri. That group had exhibited antique tractors at the fair before. They called him. He, too, felt the idea had merit and enlisted two club colleagues, Gordon Stegner and Gary Davis, to help organize the first show.

"That initial show in 1989 went well, but because it was only held one day, many fair-goers didn't see those superb tractors. We've since extended the show," Chrisenberry notes.

Thus, Larkin, Harper, and Chrisenberry, a diverse trio brought together by ancient iron, were the key players in establishing The Best of the Best antique tractor show.

Though Dianne's a city girl, her dad was an implement dealer. "He taught me how to drive a car by first learning to drive a tractor. He had a big, used Minneapolis-Moline that I drove around the back lot to get used to clutches, gears, brakes, and steering before I got in a car to drive," she says.

tional row-crop tractors—F-12, F-20, F-30, and his Regular—manufactured during the 1920s and 1930s. He isn't strictly "red-eyed," however: he also has John Deere and Allis-Chalmers models.

His current projects differ widely—a McCormick-Deering 10-20 and an Empire.

"The Empire is not well known," Chrisenberry says, "It was manufactured back East for a couple of years in the late 1940s and not many were produced. There are only 53 known to be scattered around the country, primarily in the eastern states. I bought mine in Nevada—Missouri, that is—from the fellow who bought it new in 1947.

"I do virtually all of the work when I restore a tractor unless its something like head work or sandblasting wheels."

On his business card Chrisenberry has this wording: "Convalescence home for tractors where they rust in peace." He does have a convalescence home, but his tractors really don't "rust in peace"; instead they rest in all their original glory.

Harper grew up on a 320-acre farm near Butler in Bates County, Missouri, which he now owns. Both his grandfather and father were mechanically minded and tended to be "innovative."

"My grandfather had one of the first cars in Bates County. It was a 1910 Mitchell. And," Harper said, "he was one of the first rural residents in his neighborhood to have electricity—from a Delco unit.

"My dad's first tractor was a 1917 La Crosse Happy Farmer. He had two,"

Trio of tractor buffs have helped make the Missouri State Fair a strong show for Show-Me State fans. From left are Gary Davis, Gordon Chrisenberry, and Gordon Stegner. All are members of the Chilhowee Antique Farm Machinery Collectors Organization in Missouri.

Harper said, "and he said they were the worst tractors he ever owned."

As Harper grew up he had working experience with such tractors in his extended family as the F-12, F-20, F-30, Farmall Regular, John Deere B, and a 1949 Ford 9N, the last tractor his dad owned.

Harper joined *Missouri Ruralist* in 1965 as an assistant editor. "I started collecting model tractors so I'd have something to put on the shelves in my office. I paid $24

for six models, and I sold them at a 1992 auction for $1,500," Harper said.

Harper went from models to the real thing in the 1970s. He restored his dad's 1941 John Deere B and drove and exhibited it at Bicentennial events in 1976.

His interest in old tractors continued to grow. He had heard his dad talk many times about the La Crosse Happy Farmer, and he wanted to see one. In 1984 he called the Midwest Old Threshers Reunion's office in Mt. Pleasant, Iowa,

Friends kid him that his Farmall Regular doesn't look like one. Gordon Chrisenberry's rejoinder is, "The as-it-was memory means more to me than if it were fully restored, and, besides, I have other Internationals that are fully restored." He bought this tractor as a high school freshman to start farming.

and asked the curator if she knew of any in existence.

"Sure, we have one on loan from its owner in our museum," she told Harper. She gave him the name and address of the owner and he made arrangements to meet during the Threshers Reunion. "When I got there, the owner was getting the Happy Farmer ready for the parade.

"As the parade was starting the owner said, 'Get on, you're going to drive it.' That was a great feeling," Harper recalls. "I was driving a tractor just like the one my father and grandfather drove more than 60 years earlier."

Harper has many old tractor stories to tell.

"Billy Summers, an auctioneer acquaintance, called me one day back in the 1980s and said he was having a farm sale the next day and he was going to sell an F-12. The engine was stuck, but otherwise it was a complete unit, and he wanted to know where he should start the bidding. I told him $150 should be a safe starting point," Harper says.

"Two days later Billy called again and told me to come get my $140 F-12."

That tractor, unrestored, sits in Harper's front yard and is a neighborhood landmark. Harper's wife, Jean, when giving directions to their home, says, "You can't miss us—there's a rusty tractor parked in the front yard."

Another of Harper's stories involves a 1936 Allis-Chalmers Model U. He first saw the tractor during a 1973 Missouri Cattlemen's Association tour near Rocky Comfort, Missouri, in the extreme southwestern corner of the state.

"It was sitting in the barnlot, where it had been used to blow silage," Harper said. "It hadn't been used for several years. I bought it for $125 but didn't pick it up until 1984, 11 years later. I was prompted to get it then when I noticed the former owner had filed for bankruptcy and I realized I'd better get the U before it was tied up in legal proceedings.

"In 1989 one of my neighbors who shares my interest in old tractors offered me a deal on the U. I said, 'You fix it up, and I'll pay for anything needed, and if you ever want to sell it, I get first dibs,'" Harper says.

"Now that the U is fully restored it makes a nice show tractor and we both take it to various events.

"My contribution to the sport or hobby of antique tractor collecting has been more in promoting the cause than in restoration. Maybe I'll get around to more of that later," Harper remarks.

His Tractors are Picture Perfect

Ralph Sanders loves to "shoot" classic tractors

In the crinkled family album, there are sepia-toned pictures of two-year-old Ralph Sanders toting tractors to the sand pile. When his grandmother gave him a harmonica one Christmas, he inquired of her, "Where's its wheels?" At age five, in 1938, his father traded in his last team of horses for a used Farmall F-12. "John," Ralph's Uncle Bain observed to Ralph's father, "You're going to raise three boys who won't know how to harness a team of horses." Ralph's father paused briefly and quickly snuffed that friendly criticism with, "You know, I never yoked a team of oxen and I'm getting by."

As Ralph grew up, the F-12 became his tractor. "I drove that tractor long before I could reach the clutch with my foot," remembers Ralph.

His parents and older siblings tell him that he used to go to fairs and gather up all the literature on tractors, and then take all those tractor books with him everywhere, even to bed.

Ralph's intense interest, thorough understanding, and appreciation of farm tractors carries over today and makes him the ideal photographer for the Classic Farm Tractors calendar project. During the period 1990 through 1994, Sanders photographed every one of the

He knows tractors forwards and backwards, just as he knows his profession, photography. Together, they make Ralph Sanders the ideal photographer for the Classic Farm Tractors Calendar. He took them all 1990-1994.

76 vintage, mint-condition classics that appeared on the calendars. The tractors were photographed from coast to coast and included three Canadian tractors.

For Sanders, it's a highly satisfying,

pleasurable project that gives him a warm, fuzzy feeling, sort of like a child with a new puppy.

In fact, it's probably one job assignment this professional photographer from Des Moines would do for free. "I really do enjoy it. For one thing, the calendar project has reconfirmed the strong love I have for my country and its people. These tractor folks are just like the ones we had at home growing up. They are down-to-earth, hard-working people who enjoy some of the nostalgia of the past and are preserving it," he points out.

The tractor owners are delighted that Sanders is there to photograph their pride and joy, and they cooperate fully to get "just the right background in just the right light." Sometimes the picture-taking is a piece of cake—everything works to perfection and the pictures get taken in short order. Even so, this could mean three or four hours of work: a picture can be taken in an instant; perfection takes a little longer.

For the marquee calendar photograph, he works with a large 4- x 5-view camera that requires a tripod, heavy as an anvil, plus other photographic para-phernalia. The camera's old-fashioned ground-glass back necessitates a large

The heavy, cumbersome tripod is readied, as the subject waits.

Tractor must be located exactly; here, slightly forward.

black cloth that he ducks under to compose the image and sharpen the focus on the tractor. On a hot, muggy day, there's sweat equity involved, and for that Ralph always has a large handkerchief in his hip pocket.

"Operation of the view camera fasci-nates the tractor collectors," Ralph declares. "Since I'm viewing the actual image of the tractor on the ground glass, I often let them look at their tractor 'live' under the dark cloth." Their response tickles Ralph. "But it's upside down!" they exclaim. "Well," Ralph will chuckle, "We can turn it right side up after we process the film."

Weather can be a pain. In 1993, the year of the Great Midwest Flood, a picture-taking session of an Iowa calendar tractor was set and reset five times, and on the day of the shoot things were so wet they nearly stuck the tractor.

"We try to schedule around weather patterns, but we've had to put the camera away because of rain more than once. I like bright sunlight for our shoots because we get deeper, brighter colors. I can't make the sun shine, but I try to be there when it does," Ralph smiles.

He recalls how the wet weather put

There's a photographer under there sharpening the focus.

Watching for clouds, checking the light is a constant.

Read. Aim. Click. Another tractor photo is recorded.

him behind the eight ball in 1993. "My wife, Joanne, and I did what we call a 'run and gun,' where we shot seven tractors in seven states in 10 days, driving our minivan the entire way. We shot two tractors in Illinois, and one each in Ohio, Massachusetts, Maryland, Ontario, Canada, and Michigan. We only missed one tractor due to weather. Joanne grew up on an Illinois farm just like I did, so she enjoys talking with tractor owners, too," he says.

Unpredictable weather gives Sanders a deeper appreciation of what farmers face.

"I've adopted the philosophy that if the weather isn't what you want, you have to wait, and farmers are used to that way of thinking. Patience is part of their nature."

Down-home hospitality is another trait rural folks have that Ralph encounters on his tractor trips.

"There's always coffee, and you

Sanders' farm background and familiarity with tractors make him feel right at home with tractor collectors

engines with turbochargers, intercoolers, and computer chips. It's very intricate. It will definitely take someone with some real dedication," Sanders believes.

His rural background makes his job even easier. Several family members still farm, so he's close to the land, even though he lives in the city.

His subjects have a lot in common. "They are all such willing subjects, and they're such great guys. They are all really interested, not only in what they collect, but in what other people collect. They'll help each other find parts for tractors and one of their greatest joys, I think, is being able to parade these restored tractors in their local parades rather than taking them to competition.

Our photographer is in the clover—for the right angle.

usually wind up eating a meal with them if it's that time of day," he says. "I had one couple insist that I take some bedding plants home with me, so our tomato crop one year came from Harold Glaus in Nashville, Tennessee. They were excellent plants." (For more information on Harold Glaus, see the chapter entitled It Only Takes One on page 72.)

Sanders says there is one common concern among these collectors.

"The question I hear most often is, 'Who's going to preserve these modern-day tractors?' You've got enormous diesel

Of course, they also show them at state fairs and tractor shows, but I think they would rather do it for their own personal pleasure and their town rather than to compete.

"Almost without exception, they'll collect something they knew and liked as youngsters. Their fathers or grandfathers or somebody they knew had that tractor. I, too, would like to have an F-12 Farmall sometime—that's one of the $\frac{1}{16}$ scale models I have. I've eight or nine models and they're all meaningful to me, especially the rusty IH crawler I had as a boy. Many collectors may expand or start collecting other models, but it usually starts with buying a tractor they grew up with," notes Ralph.

He's made fast friends "shooting"

these subjects and their machines.

"We keep in contact with quite a few of them. As a matter of fact, one family, the Steve Rosenbooms (see page 48), have stopped twice to visit us in Des Moines while attending the state basketball tournament."

Sanders, 61, and Joanne, have seven

Ralph Sanders flashes a smile after a successful shoot.

"Hold it right there, don't blink, and now say cheese."

grown children and six grandchildren. Two of their boys, Scott and Rich, are involved with the family business, Sanders Photographics, and keep it humming when Ralph and Joanne are on the road.

Joanne had never been

directly involved with the calendar shoots until 1993, but then the travel bug bit and now she's raring to go with Ralph on every road trip.

Ralph has been chasing tractors and other farm equipment to photograph for more than 20 years. In 1974, after 16 years as an agricultural journalist, he and Joanne established Sanders Photographics in Des Moines, Iowa, to do advertising and commercial photography on a free-lance basis.

"Twenty years without a steady job has been more fun than we ever imagined," Ralph says, "and we're planning for at least 20 more."

Decking Tractors With Christmas Cheer

Sunnyside, Washington, salutes agriculture with an annual evening Christmas parade of lighted tractors and equipment

Suppose you lived in the heartland of a diversified agricultural region, and your entire community was tied to agriculture. How do you celebrate this agricultural heritage, get everyone involved, and have fun doing it?

You could start with a parade of antique farm tractors. Then add Christmas lights, and make this an annual event during the holiday season. Let Christmas-themed floats, high school bands, modern farm equipment, and late model tractors join in and what do you have?

You have the Sunnyside (Washington) Country Christmas Lighted Farm Implement Parade, and a dream of Bob Hadeen's come true.

It was Bob, a long-time resident of the Sunnyside community, who came up with the idea of a parade featuring lighted farm tractors and farm implements of every description. "I felt this kind of parade would be even more unique than lighted Christmas boat parades held in many communities located near bodies of water," he explains.

His philosophy on life is refreshing. "When asked where I get my ideas, my reply is, 'I manage by wandering

around.' It's not necessary to reinvent the mousetrap, only improve on it. When one views a happening in some other area that is successful, ask yourself, Can we use it in our community and if so, how can we adapt it? Most

people view life's happenings without really seeing them," Bob feels.

It took a while to sell the idea of the parade, because some thought it was crazy. However, in 1989, the Sunnyside Country Christmas Lighted Farm Imple-

Deere decorations, consisting of candy canes and Christmas lights, are attached to Frank Schilperoort's 1937 John Deere L, by Frank (wearing hat) and Bob Hadeen, preparing it for the Sunnyside lighted tractor parade. Hadeen first proposed the parade idea which has become a tradition in this south central Washington city.

A gigantic grape harvester featured a swinging Santa Claus (Tom Colton) up front, all courtesy of Rattray Farms.

ment Parade became a reality, lighting up south central Washington like the aurora borealis light up Alaska.

The parade lives up to its mission: "To capture the spirit of Christmas in Sunnyside's rural environment." Sunnyside, Inc., Community Economic Development and the Sunnyside Chamber of Commerce have been the parade's sponsors, and now the event ranks as one of the major visitor and tourist attractions in central Washington during the Christmas season.

The parade is held on the first Friday following Thanksgiving weekend—after dark, naturally. Some 75 entries in four categories compete for prizes, and every entry is loaded with Christmas lights.

Officially, the four parade divisions consist of: old farm equipment and tractors; commercial farm equipment and tractors; Christmas floats; and modern farm equipment and tractors.

Sunnyside is situated in the Yakima Valley, one of the most diversified agricultural areas in the world, producing everything from apples, asparagus, hay, hops, milk, mint, fruit, wheat, and corn (some 40 crops in all). At the parade, therefore, one is likely to see combines, harrowbeds, plows, tillers, spreaders, choppers, swathers, grape pickers, and many more farm machines.

All pay tribute to Sunnyside's agricultural heritage, past, present, and future.

The fifth annual parade, in 1993, received national recognition when it was featured on the CBS News Sunday Morning Show with Charles Kurault. "We received hundreds of positive comments from across the country after the program aired," notes Dave Fonfara, director of Sunnyside, Inc.

Anyone who wants to enter a tractor in the parade must follow the first rule to the letter: "All entries must be completely lighted and the lighting must represent our 'Country Christmas Theme.' " That rule applies to every tractor, piece of equipment, truck, and float.

"Stringing up a few lights on a tractor will not be accepted," the rules firmly state.

Parade regulars such as Todd Monroe spend a full day decking their antique tractors with Christmas lights. His 1948 John Deere B and the plow it pulls are all decorated with hundreds of Christmas lights, even the wheels.

What a wonderful way to honor one's agricultural roots. What a wonderful way to get into the Christmas spirit.

What's this, another Santa Claus? We'll have to ask Kyle Shinn for an explanation. That's his 1949 John Deere D.

(Lower right) Darrell Wheeler wheeled along with his 1953 John Deere M pulling a trailer loaded with a Christmas tree, teddy bear and presents.

New tractors like this Ford 5030 participated in the Sunnyside, Washington Parade, alongside antique models.

You'll never guess what this John Deere 4255 is pulling. Give up? It's an asparagus bedding tiller.

Hart-Parr Tractors Are Dear to His Heart

*With nearly 50 machines under renovation, the Show-Me state's
Jim West has lots to show*

*Jim West wants a tractor that will run, even if
he does here with a Hart-Parr 12-24, built in the
late 1920s.*

On a shelf in the kitchen of Jim and Mary Lou West sits a plaque that reads, "The main difference in men and boys is the price of their toys." Ranged along the shelf on either side of the motto are toy tractors, Oliver Hart-Parrs of virtually every make and model built.

These miniature models, however, hardly represent the price of Jim West's toys. The LaPlata, Missouri, farmer has nearly 50 full-sized Olivers and Hart-Parrs, from the behemoth two-bangers of the early 1900s to relatively rare Oliver lawn-and-garden models. In fact, West has built a 32- x 60-foot display shed for his collection.

"I can't very well take more than one or two old tractors to a show," says West. "I wanted a place to exhibit them all together, for other people who have Hart-Parrs in their hearts."

Jim West took Hart-Parr tractors to his heart about half a century ago, when he was a boy growing up on his family's crop and vegetable farm in New Jersey.

"In 1936, my father bought a new Oliver 70 row-crop for $700," recalls West. "I started cultivating potatoes with that tractor when I was seven years old."

The Oliver 70 now is the centerpiece of West's collection of restored tractors. Jim and Mary Lou moved to Missouri in 1972, and shortly afterward went back to New Jersey to bring the Oliver home with them.

"With that tractor, my Dad went contrary to what most farmers did," West adds. "The Model 70 came out on rubber originally, but he switched it to steel tip-toe. In fact, he had both: rubber and steel. He also put a wide front-end on it later on."

In cultivating truck crops, especially potatoes, steel "tip-toe" wheels cut the vines rather than mash them, West notes. With a potato vine cut cleanly, the plant healed quickly. There was less opportunity for disease and insect damage.

"Olivers and Hart-Parrs were popular in New Jersey, both row-crop and orchard models," says West. "For three years in the early 1930s, Oliver Hart-Parr sold more tractors in New Jersey than in all the rest of the country put together."

Which was not a massive movement of machinery, at that. During those Depression-era years, machinery manufacturers fell on hard times because

West restored this Oliver Super 44, and has a replica of it as part of his collection of toy tractors.

His 1936 Oliver Hart-Parr, an orchard model, won first prize as the "Best on Steel" at the Missouri State Fair's Antique Tractor Show in 1992.

farmers had no money to buy new equipment. Oliver Hart-Parr only sold 20 of their 18-28 models in 1932. West has one.

West recalls some history of the company that made his favorite tractor: Hart-Parr Company was the first U.S. manufacturer to successfully market gasoline-engine tractors. Charles W. Hart and Charles H. Parr organized the firm at Madison, Wisconsin, in 1897, then moved the operation to Charles City, Iowa. In 1907, Hart-Parr introduced its Model 30-60, with two horizontal, fore-and-aft cylinders. This tractor, dubbed the "Old Reliable," was a best seller. More than 400 units were sold during the initial model year.

In 1929, Hart-Parr merged with Oliver Chilled Plow Company and the Nichols & Shepard Company (builders of grain threshers) to form the Oliver Farm Equipment Corporation. "My interest in Oliver Hart-Parr tractors is the nostalgic tie to the kind of tractors we used on my family's farm," says West. "I first got interested in Hart-Parr because of my family's history with that equipment. But there aren't many tractors left of some of those old models. I want to keep enough of them in good condition so people can see what they looked like."

And of the dozen or so tractors Jim has completely restored, what they look like is brand new. West's inherited 1936 Model 70 row-crop won third place at the Missouri State Fair's Best of the Best Antique and Classic Tractor show in

1991. A year later, West took the Best on Steel category with his 1936 Oliver Hart-Parr orchard Model 70.

"This tractor is serial number 287," West points out, of his prize-winning Model 70 standard. "It's the 287th standard built. But Oliver didn't keep separate serial numbers for orchard and standard models. They are all numbered sequentially, so this is the 287th Model 70 built by the company."

As a member of the Oliver Hart-Parr Collectors' Association, based in Charles City, Iowa, West keeps up with what other Oliver restorers are doing. The association also is a good source of

hard-to-find parts for those tractors in his own shop.

"Some of the old tractors we get are in pretty sad shape," he says. "We have to do a lot of work on them, especially on the sheet metal. But it's worth it, to me."

Each winter, after maintenance and repairs are finished on his farm equipment, you'll find Jim West hard at work restoring another antique tractor.

"I try to get at least one old tractor done each winter, but that depends on how much work is involved," West concludes. "This is an expensive hobby in [terms of] both hours and dollars."

Those old Oliver Hart-Parrs in Jim

West's shed, lined up like military cadets on inspection day, are not toys for boys. These are the playthings of a devoted craftsman who wants to preserve bits of the past—in mint condition.

This 1936 Oliver 70 still carries the Hart-Parr name. West learned to drive this tractor as a seven-year-old on his family's vegetable farm in New Jersey.

In a Class All Its Own

Donald Kingen is an owner of the famous Minneapolis-Moline UDLX, a tractor far ahead of its time

The year was 1938. Franklin Delano Roosevelt was in the White House. A large number of Americans still lived on the land, and many of them still farmed with horses. Things were tough and money was tight.

However, tractor manufacturers—and there were a large number still in the business—knew tractors would eventually prevail, and that change was in the air. That year 51 models of nine makes not previously listed were introduced to the tractor market.

John Deere added the G, a large three-plow tractor; and other Deere models got a facelift or "restyling." Oliver rolled out its Row Crop 80 series, while Allis-Chalmers announced its Model B. The Avery group in Peoria surprised everyone with a straddle-row model named the Ro-Trak, that could be easily changed from a tricycle type to a standard tread.

Massey-Harris came forth with the 101 and Twin Power, while the F-14 joined International Harvester's famous Farmall series. Eagle added the 6C model to its line, and Caterpillar announced its small D-2 for farm work.

But the best was saved for last.

At a farm tractor show that fall, Minneapolis-Moline jarred the industry with its introduction of the "Comfortractor," the UDLX or U Deluxe, a tractor unlike any on the market anywhere.

Don Kingen, of McCordsville, Indiana, is the proud owner of a 1938 Minneapolis-Moline UDLX Comfortractor, one of the most sought-after classic tractors around.

This tractor came with a factory-built cab (first of its kind), though you could buy it with or without cab. It had five forward speeds, from a crawl to 40 miles per hour—really speedy for a tractor in 1938.

"It was the only year they were built," says Donald Kingen, of McCordsville, Indiana. "They made 150 of them—that was total production—all in 1938."

Because only a handful were made in the first place, and only a few remain, this avant-garde Minneapolis-Moline model has become one of the most collectible of all classics.

"I happened to see an auction advertised in *Gas Engine Magazine*," Kingen recalls, "and a UDLX was listed." An elderly gentleman in Iowa had passed away and the tractor would be sold in the estate sale. It had not been restored.

"You always dream about owning one, and you don't see one for sale very often. We were right in the middle of corn planting. You know you've got it bad when you take off and drive to Iowa and leave the boys with most of the corn to plant," Kingen muses.

He outbid others for the UDLX in Iowa and brought it back to Indiana for restoration.

This 1938 Minneapolis-Moline UDLX (ultra deluxe) came from the factory equipped with a cab, radio, cigar lighter, heater, hub caps, paneling, a horn, headlights, even a license plate holder. It was capable of whizzing down a country road or highway at 40 miles per hour. The idea was to plow all day, unhitch, and drive to town that night. Some say only 125 were built, others say 150. Either way, it is a highly desirable collectible.

There were some missing parts, but tractor buffs have their own network, their very own information super-highway.

"The Comfortractor came with a heavy-duty front bumper, but mine was missing. I heard about a fellow in Saskatchewan, Canada, 300 miles north of the Montana border, who had four of these tractors, trying to make one good one, and he had an extra bumper. So, by word of mouth, I ran onto it. Sure enough, he shipped it down here to me.

"It's hard to believe, but they used one of these tractors on a mail route there in Canada until just a few years ago. Guess they would go in the snow," says Kingen.

His UDLX has the original grille, and he bought a handsome piece of M-M insignia for the grille from Roger Mohr, a long-time collector of Minneapolis-Moline tractors and memorabilia, who also owns a UDLX. The nifty, original M-M hood ornament was still on the tractor when Don got it.

Would this tractor really go 40 miles per hour?

"I had it up to 40 once on the road in front of the house, but that was too fast . . . you get all the iron moving," Don says, shaking his head. "But that was the whole idea. You were supposed to plow all day, unhook, and go to town that night. You didn't need an automobile."

There are numerous other features of this unique antique.

"It had twin headlights, with high and low beams, plus a brake light and license plate holder. Luxuries in the cab included

Streamlined and looking more like a car than a tractor, the price of the UDLX was about $1,800, which put it out of reach of most Depression farmers. Its cost was twice that of other similar-size tractors of its day.

a radio, cigar lighter, ash tray, and paneled dash," notes Kingen. There was also a knick knack compartment, and, are you ready for this? An electric horn.

"The tractor sold for around $1,800 or a little more, and that was the catch. It was right after the Depression, money was still scarce, and that price was twice what a John Deere or Farmall cost. And remember, tractors were still competing with horses, too," he says.

The Hoosier State collector got lucky and was able to purchase a set of 32-inch rear tires. "You can buy 30-inch or 34-inch, but no 32-inch tires," he says.

He bought the tires at a county highway auction in eastern Indiana. "They had two sets of them and I was so excited I was shaking like a leaf in the fall. I bought this one set and I had them in the pickup and I was gone. Before dark that night they were on the UDLX. I didn't give much for them and I thought I'd better go before they changed their mind. I should have bought the other set, too, but I was just too excited," he smiles.

It received lots of press, but it didn't win over farmers. It also had windshield wipers, a defroster, a foot accelerator, and used the same high compression engine developed by Minneapolis-Moline for the Model U series that set new standards for the industry.

The high-compression, 283-cubic inch, 6-cylinder engine was rated at 35 horsepower. This Minneapolis-Moline engine was to set new standards for the industry, as well. (In the spring of 1935, M-M sold the first high-compression tractor, the KTA, known for power and fuel economy.)

"They called it a straight block engine, and it was the basic motor used up to 1954. The UDLX engine even had a pressure cooling system; again, ahead of its

time. It had an oil bath air cleaner, even a fuel gauge in the gas tank," Kingen notes.

The company promoted the UDLX with gusto, spending lots on promotions, advertising, and dealer demonstrations. "Miss Minnie Moline," a young lady dressed in fancy bib overalls, took readers on the tour of the new tractor in a colorful brochure with the headline, "Hats off to a greater modern tractor by the modern tractor pioneers."

At least 30 years ahead of its time, it was a generation later before the totally enclosed tractor cab was reintroduced.

In the cab of the UDLX, the driver rode in comfort with cushioned seats for two. Fully protected from dust and the elements, the cab was complete with safety glass, and afforded amazingly good vision front, rear, and sides.

On the dash was a speedometer, ampmeter, and oil pressure and water temperature gauges. There was a rear view mirror with clock. Fenders and hubcaps dressed up the exterior.

The variable speed governor was controlled by a pedal, like a footfeed. The brakes were Bendix; they, and the clutch, were foot operated. And like an auto, there were windshield wipers. Talk about being loaded. This tractor had all the extras—in 1938!

Some have compared the Comfortractor with the Tucker automobile that was introduced following World War II. It, too, was stylish, streamlined, and customized—and far ahead of its time.

Kingen shares his UDLX with others at numerous antique tractor gatherings, and such major events as the Indiana State Fair and Farm Progress Show. Many people have never seen this tractor, and stand and stare in awe. Others comment on its resemblance to a luxury car and its high speed capability.

Is it any wonder collectors such as Donald Kingen, regardless of how many vintage tractors they may own, point to the 1938 Minneapolis-Moline UDLX as their pride and joy?

About the Author

"He's just an itinerant journalist," one might think when first looking at John Harvey's resume.

It's true that John has moved about during his career as an agricultural journalist. But, unlike an itinerant who might move on a whim, each of John's moves was given serious thought and careful discussion beforehand with family, friends, and fellow professionals.

As a result, each of his career moves was a step forward to a more challenging position and one where he could develop still other professional skills.

His career in journalism started as soon as he completed his degree in agricultural journalism at the University of Missouri in 1957. And, like many journalists, John started his career as a writer/photographer for a small weekly newspaper—the *Savannah Reporter*—in his home town of Savannah, Missouri.

Following his thorough indoctrination into the journalism profession, John took his first step up the career ladder. He joined the staff of *Missouri Farmer*, the monthly magazine of the Missouri Farmers Association in Columbia, Missouri, as field editor.

After seasoning at the state publication level, John moved on to become associate editor of *Successful Farming*, in Des Moines, a general farm magazine with a national circulation.

Having experienced and mastered writing and photography at the local, state, and national levels, John's career changed direction when he joined Reiman Associates in Milwaukee, in 1970. There, he became involved in agricultural public relations.

Next, John tasted public service in accepting a special U.S. Department of Agriculture assignment. His task was to help create and write the Department's Bicentennial Yearbook of Agriculture, "The Face of Rural America," published in 1976. In this assignment, John worked with leading photographers across the United States in depicting U.S. agriculture with 335 photographs.

Moving on, John joined the editorial staff of *Farm Journal*, Philadelphia, the largest general farm publication in the United States.

Another carefully considered step was to DuPont Agricultural Products in Wilmington, Delaware, where for 16 years he promoted U.S. agriculture and America's farmers, and their interests.

At DuPont in 1990, he created the Classic Farm Tractors Calendar, to promote the newly developed DuPont Classic herbicide. The calendar featured restored farm tractors—most 50 years old or older—one for each month, from across the United States. The calendar was an instant success. Thousands were given to DuPont customers and thousands were sold. It has become one of the most popular calendars in the country.

His most recent career move came in 1993, when upon leaving DuPont, he established John Harvey Communications, serving as a communications and public relations consultant, with emphasis on America's agricultural heritage.

In 1994, he introduced Classic Tractor Playing Cards, featuring a vintage tractor on all 52 cards, another first.

During his 37-year career, John has won many of the major writing, photographic, and public relations awards in agricultural journalism profession. Included is the revered Oscar in Agriculture, honoring excellence in the agricultural reporting, and the Agricultural Relations Council's highest honor, the ARC Founders' Award.

John has won other recognition. The National Association of Farm Broadcasters, the American Soybean Association, and his alma mater, the University of Missouri, have presented him with meritorious service awards.

Though he's happy in his Wilmington, Delaware surroundings—his home for 18 years—he maintains a strong affinity with his native Midwest. (He laughingly claims he can't get a breaded pork tenderloin sandwich, or a Cherry Mash candy bar—two of his favorite foods—on the East Coast!)

Harvey has his ever-supportive wife of 33 years, Carol, and his daughter and two sons nearby to provide strong personal encouragement.

"John Harvey is a gifted writer, a devoted husband and father, and, deep in his heart, a true lover of the land," wife Carol declares.

— By Dick Lee, former
Agricultural Editor,
University of Missouri-Columbia